Springer Series in Optical Sciences

Volume 173

Founded by

H. K. V. Lotsch

Editor-in-Chief

W. T. Rhodes

For further volumes:
http://www.springer.com/series/624

Springer Series in Optical Sciences

The Springer Series in Optical Sciences, under the leadership of Editor-in-Chief William T. Rhodes, Georgia Institute of Technology, USA, provides an expanding selection of research monographs in all major areas of optics: lasers and quantum optics, ultrafast phenomena, optical spectroscopy techniques, optoelectronics, quantum information, information optics, applied laser technology, industrial applications, and other topics of contemporary interest.

With this broad coverage of topics, the series is of use to all research scientists and engineers who need up-to-date reference books.

The editors encourage prospective authors to correspond with them in advance of submitting a manuscript. Submission of manuscripts should be made to the Editor-in-Chief or one of the Editors. See also www.springer.com/series/624

Matteo Perenzoni · Douglas J. Paul
Editors

Physics and Applications of Terahertz Radiation

 Springer

Editors
Matteo Perenzoni
Center for Materials and Microsystems
Fondazione Bruno Kessler
Trento
Italy

Douglas J. Paul
School of Engineering
University of Glasgow
Glasgow
UK

ISSN 0342-4111   ISSN 1556-1534   (electronic)
ISBN 978-94-007-3836-2   ISBN 978-94-007-3837-9   (eBook)
DOI 10.1007/978-94-007-3837-9
Springer Dordrecht Heidelberg New York London

Library of Congress Control Number: 2013947053

Printed on acid-free paper

Springer is part of Springer Science+Business Media (www.springer.com)

Preface

The "Physics and Applications of THz Radiation" book will cover the latest advances in the techniques employed to manage the THz radiation and its potential uses. It has been subdivided into three sections: THz Detectors, THz Sources, Systems and Applications. These three sections will allow the reader to be introduced in a logical way to the physics problems of sensing and generation of the terahertz radiation, the implementation of these devices into systems including other components, and finally the exploitation of the equipment for real applications in some different field. On the other hand, any of the sections and chapters can be individually addressed in order to deepen the understanding of a single topic without the need to read the whole book.

The THz Detectors section will address the latest developments in detection devices based on three different physical principles: photodetection, thermal power detection and rectification. Quantum well devices for photodetection will be introduced by first defining the physics of heterostructures, which is also used for infrared detection, then issues related to the lower photon energy of the terahertz range will be addressed and solutions proposed. Bolometric detectors will be explored in their manyfold aspects, from the basics that are general for each device to more specific features, in particular the characteristics and coupling needed for terahertz detection. Devices based on field-effect transistors rectification will be introduced starting from the underlying physics, dividing it in the resonant and non-resonant cases, and then more advanced and specific characteristics will be analyzed.

The THz Sources section will describe three completely different generation methods, operating in three separate scales: quantum cascade lasers, free electron lasers, and nonlinear optical generation. Lasers based on the quantum cascade effects will be explored starting from the basics of lasing in multiple quantum well structures, and continuing in exploring the problems of terahertz generation with a glance towards future developments. Sources based on the use of relativistic electrons will be described with an eye of the basic physics but also with the Frascati free-electron laser experience and the efforts of scaling down the size of the equipment. Nonlinear optics generation will focus on the possibility to obtain compact tuneable frequency sources, from the principle to complete source systems with intra-cavity OPO.

The Systems and Applications section will take care of introducing many of the aspects needed to move from a device to an equipment perspective: control of terahertz radiation, its use in imaging or in spectroscopy, security applications and potential safety issues. Control of the terahertz radiation is addressed in particular for what concerns metamaterials, and their ability to work on the basic properties of the waves will be described. Terahertz spectroscopy will be introduced for what concerns basic equipment, then the focus will be moved onto the measurement problem with the purpose of improving the quality and reliability of the extracted sample material characteristics. Security applications, which often are addressed as the most promising for terahertz, are described in detail, starting from the requirements to the possible implementations and comparison with the near millimeter waves region. The safety issues part, finally, gives an updated look of a field that still shows controversy and explains the current results of recent studies on interaction of terahertz waves with living tissues.

Contents

Part I
THz Detectors

Chapter 1
Quantum Well Photodetectors

Fabrizio Castellano

Abstract The progress in semiconductor optoelectronics has now reached a level where commercial systems for generation and detection of radiation over the whole infrared spectrum are available. At lower frequencies, semiconductors dominate the electronics and microwave world since ever. Semiconductors are then expected to play a fundamental role in THz science also, this spectral range being the bridge between the two worlds. The success of semiconductor infrared detectors has been made possible by the development of the quantum well, the fundamental building block of all bandgap engineered structures. Here in the first sections the fundamental physical aspects of semiconductor quantum wells and intersubband transitions are resumed, along with a discussion about how they are used in quantum well infrared and THz detectors. A review of the current progress on the subject can be found in the last sections, including a mention to quantum dot detectors, which appear to be the future of semiconductor infrared detectors.

Keywords Quantum well · Semiconductor · Infrared detector · Terahertz detector · Quantum dot · Intersubband transitions · Energy gap

Infrared detection and imaging is a very broad technological field encompassing a variety of solutions. In the field of semiconductors, HgCdTe interband photoconductive detectors are the most common commercial solution for single pixel devices, while quantum well photodetectors are a competing technology mostly based on the mature GaAs/AlGaAs material system. They provide some advantages with respect to HgCdTe detectors, mainly when large area focal plane arrays or high speed operation is needed. While pushing interband detectors far in the infrared, beyond 25 μm, is not feasible, quantum wells still provide a viable choice thanks to their ability to scale the operation frequency by geometry rather than material composition.

F. Castellano (✉)
CNR Istituto Nanoscienze, Piazza S. Silvestro 12, 56124 Pisa, Italy
e-mail: fabrizio.castellano@nano.cnr.it

M. Perenzoni and D. J. Paul (eds.), *Physics and Applications of Terahertz Radiation*,
Springer Series in Optical Sciences 173, DOI: 10.1007/978-94-007-3837-9_1,
© Springer Science+Business Media Dordrecht 2014

The THz range is the latest addition to the list of technologically important frequency bands, and many different technologies are trying to access it either from the optics or the electronics side. Historically, semiconductor optoelectronics has started in the visible and near infrared and then evolved toward longer (and shorter) wavelengths. On the long wavelength side, a hard wall is encountered when polar optical phonons come into play, between 25 and 40 μm wavelength. Beyond that, the THz region conventionally starts, extending down to 1 THz frequency (300 μm wavelength).

This chapter resumes the fundamental physics of quantum well detectors, with a focus on the issues that are more relevant to THz science. Anyway, since quantum well technology has profound roots in the mid-infrared, and the device physics is pretty much independent of the operation frequency, it would be an unnatural choice to limit the discussion to devices operating strictly in 1–10 THz range. For this reason, some references to shorter wavelength devices are included when relevant.

1.1 Fundamentals of Photoconductors and Photodiodes

In order to establish nomenclature and notation, a review of the basics of infrared photodetection follows, recalling the fundamental quantities and formulas that apply to any kind of infrared detector. A broader discussion can be found in [1].

1.1.1 Detector Figures of Merit

A photodetector is any device that can produce an electrical signal proportional to the amount of light that is incident on it. Depending on the type of electrical signal produced, photodetectors can be divided in photoconductors and photodiodes. Photoconductors can be thought as resistors whose resistance varies when exposed to light. These detectors are usually biased at a fixed voltage and the current flowing though them (photocurrent) is the detection signal. Photodiodes on the other hand generate a voltage between their contacts (photovoltage) as a response to a light input, and this can be directly measured without the need of external bias. Since photoconductive detectors are by far the most developed for infrared sensing applications, the following discussion will focus mainly on them.

Given an opaque object at temperature T_{obj}, the corresponding optical power spectral density incident on a detector of area A can be calculated from Planck's radiation law as

$$\frac{dP}{dv} = \frac{A}{4F^2 + 1} \frac{2\pi h v^3}{c^2 \left(e^{\frac{hv}{k_B T_{\text{obj}}}} - 1 \right)} \tag{1.1}$$

where F is the ratio between the focal length of the last optical element and its diameter. The total power P can be obtained by integrating the power spectral density over the optics bandwidth. An ideal linear photoconductive detector will produce a current $I(P) = \mathscr{R}P$, where $\mathscr{R}$ is called the responsivity of the detector.

If every photon that strikes the detector has a probability η of being absorbed (internal quantum efficiency) and for every absorbed photon there are g electrons injected in the circuit (photoconductive gain) then the responsivity can be immediately calculated as

$$\mathscr{R} = \frac{q\eta}{h\nu}g. \tag{1.2}$$

by using the fact that to an optical power P there corresponds a photon flux $\frac{P}{h\nu}$. Photoconductive gain and quantum efficiency are fundamental parameters of the photoconductors: they work together to produce the responsivity of the detector, which in turn determines the amount of current that can be expected for a given optical power.

In real imaging systems the total optical power incident on the detector is actually the sum of two components, a signal power $P_s(T_{\text{obj}})$, coming from the object to be imaged, and a background power $P_b(T_b)$, coming from the scene behind the object and the optics, that together form a blackbody background at a temperature T_b. Thus the total current through the photoconductor can be written as

$$I = \mathscr{R}(P_s(T_{\text{obj}}) + P_b(T_b)) \tag{1.3}$$

Finally, in the case of a non-ideal photoconductor, when a bias voltage is applied, a dark current $I_d(T_{\text{det}})$ is flowing also in the absence of an optical signal, and this current depends on the detector temperature T_{det}. The total current is then

$$I = I_s + I_b + I_d = \mathscr{R}P_s(T_{\text{obj}}) + \mathscr{R}P_b(T_b) + I_d(T_{\text{det}}). \tag{1.4}$$

The above general relation can be easily written also for photodiodes, by simply turning currents into voltages.

Another fundamental aspect of photodetectors is the noise that characterizes their operation, which gives a lower limit to the amount of light that can be detected. For an ideal photodiode exposed to blackbody radiation, the only source of noise is coming from the randomness of the photon flux itself, giving rise to the so-called shot noise in the photocurrent. In the case of photoconductors, the flow of current is additionally characterized by the so-called generation-recombination noise due to unwanted electrons being captured or excited into the circuit because of thermal effects. All these noise components sum up to a noise current $i_n(P)$. Details of how an expression for the noise currents can be derived are not provided here, as they are extensively detailed in [1].

A more practical figure of merit that measures the noise in a photodetector is the noise equivalent power (NEP). This is defined as the signal power needed to obtain a signal to noise ratio of 1, and it is thus defined by the condition

$$i_n(NEP) = I_s(NEP) \tag{1.5}$$

Depending on the particular working conditions of the detector, one can define various NEP values for a photoconductor.

- In the case of signal-noise limited detection (SL), the signal is so strong ($I_s \gg I_b, I_d$) that the assumption $I \approx I_s$ holds in Eq. (1.4). This is the case when the object to be imaged is very hot or take up most of the field of view.
- In background-limited (BL) detection, the dominating current component is the one coming from the optical background ($I_b \gg I_s, I_d$), so that $I \approx I_b$ in Eq. (1.4). This typically happens when small objects are imaged over a warm background.
- Finally if the background is very cold, the dark-current dominates, so that $I \approx I_d$ in Eq. (1.4). This can happen in astronomy applications when stars are imaged on the dark cosmic background. Table 1.1 summarizes the expressions for the NEP obtained from condition Eq. (1.5) in the various detection cases for both ideal photoconductors and photodiodes.

Another commonly used important figure of merit for detectors is the detectivity D, defined as the inverse of the NEP. Since the NEP of a detector always depends on the measurement bandwidth Δf and, in the case of background limited detection, on the detector surface, the specific detectivity

$$D^* = \frac{\sqrt{A\Delta f}}{NEP} \tag{1.6}$$

is commonly used to characterize the noise properties of photodetectors. It is measured in Jones, with $1\,\text{Jones} = 1\,\text{cm}\,\text{Hz}^{1/2}/\text{W}$.

It is apparent from Table 1.1 that the noise performances of photovoltaic detectors (photodiodes) are always better than those of photoconductors. This is because in the ideal photovoltaic detector the dark current is zero and the only source of noise is the shot noise coming from the noise already present in the thermal photon radiation. In the case of a photoconductor, the additional noise comes from the fact that a bias has to be applied in order to have a photocurrent, and this current is intrinsically subject to generation-recombination noise.

Since the dark current in a biased detector typically depends exponentially on the detector temperature, there is a sharp transition from the background limited detection regime and the dark-current limited regime. The background limited infrared photodetection temperature (T_{BLIP}) is the temperature of the detector at which this

Table 1.1 Summary of ideal photoconductor (PC) and photodiode (PD) NEP expressions

	Condition	PC	PD
NEP_{SL}	$I_s \gg I_b, I_d$	$\frac{4h\nu\Delta f}{\eta}$	$\frac{2h\nu\Delta f}{\eta}$
NEP_{BL}	$I_b \gg I_s, I_d$	$\sqrt{\frac{4h\nu P_b\Delta f}{\eta}}$	$\sqrt{\frac{2h\nu P_b\Delta f}{\eta}}$
NEP_{DL}	$I_d \gg I_s, I_b$	$\sqrt{4qgI_d\Delta f}$	0

transition happens. Its value is defined as the temperature at which the dark current (noise) equals the background current (noise). For temperatures below the T_{BLIP} the detector operates as if there is no dark current, since its noise contribution is negligible with respect to the noise induced by the optical background, and the attainable signal to noise ratio depends only on the temperature of the optical setup around the detector (lenses, mirrors, room radiation). Above the T_{BLIP} the signal to noise ratio deteriorates exponentially as the temperature of the detector is raised. Thus, the T_{BLIP} sets the practical operating temperature range of a detector and it determines its technological and commercial exploitation potential. Typical T_{BLIP} values range from 150 K for mid-infrared quantum well detectors, which is accessible with liquid nitrogen, to around 10 K for THz devices, which requires liquid helium.

1.1.2 THz Range Specific Issues

In the field of quantum well detectors efforts are always aimed at increasing the T_{BLIP} to be as high as possible, ideally to room temperature values, or at least in the temperature ranges where it is possible to employ compact thermoelectric coolers rather than cryogenic gases. At the moment this is the single major limitation to the widespread exploitation of quantum well detectors for passive long-wavelength infrared and THz imaging.

Quantum well detectors base their operation on electron confinement in nanostructures. This detection means small confinement energies, in the 4–20 meV range for the THz, a value that becomes quickly comparable with the average thermal energy of the electrons as the temperature raises. As a result, semiconductor THz detectors exhibit T_{BLIP} values typically below the temperature of evaporating liquid nitrogen (77 K), meaning that liquid helium cooling is required to achieve background limited detection. This fact limits the potential applications of THz quantum well detectors to high-budget systems, such as military, astronomy, and research apparatuses.

Another major problem, affecting all types of THz detectors, is the intrinsically low power levels of thermal radiation emitted in the THz band. From equation Eq. (1.1) it is in fact apparent that such power scales as the cube of the frequency. This means that for the same object at a temperature T, the emitted power at a wavelength of 100 µm (3 THz) is 1000 times smaller than the power emitted in the 10 µm (30 THz) range. This translates in a 1000 times bigger collector optics if the same detected power levels are required.

1.2 Semiconductor Quantum Wells

The term "quantum well" refers to systems where the motion of electrons has been constrained in one direction, resulting in a two-dimensional motion of the particles and the appearance of quantized energy states.

Such systems can be realized in semiconductors by engineering the conduction band profile of the material along one direction in order to produce potential variations on the nanometer scale. This is made possible by semiconductor crystal growth technology such as molecular beam epitaxy (MBE), which allows the fabrication of semiconductor structures with atomic layer control. During MBE growth a beam of evaporated semiconductor atoms is targeted toward a heated semiconductor crystal substrate. As the atoms reach the hot surface they bind to the existing atoms creating additional crystal layers one after the other, reproducing the substrate's crystal structure. The great advantage of this technique lies in the fact that the chemical composition of the layers need not be exactly the same as the substrate and it can be varied by changing the relative intensities of the molecular beams. This allows for different semiconductors to be grown on the same substrate, sharing the same crystal structure and together forming a new layered crystal (heterostructure) whose electrical and optical properties are determined by the thickness of the layers. The most common material system of choice for quantum well detectors is the $GaAs/Al_xGa_{1-x}As$ material system grown on a GaAs substrate, which can reliably provide confinement potentials up to few hundred meV.

Regarding device fabrication the technology is based on established lithography/etching/metallization techniques inherited from the electronics world. The lithographic resolution required is never stringent, being always larger than $1\,\mu m$, allowing for the use of standard UV lithography. The availability of already developed high quality materials and processing techniques is what boosted research in quantum well science in its infancy.

It is not our goal to give a detailed explanation of fundamental physics of semiconductor quantum wells as many books from authoritative authors exist on the subject [2, 3]. The following resumes the fundamental aspects of quantum well physics, underlining the concepts that are most important for THz detectors.

1.2.1 Quantum Wells

The quantum well is the building block of infrared optoelectronic devices. In its most basic form it consists of a layer of narrow-gap material sandwiched between two large-gap layers. The resulting conduction band profile exhibits a square well shape whose width is determined by the thickness of the narrow gap material, and whose depth is dictated by the conduction band discontinuity between the barrier and well material.

When the quantum well thickness is on the nm scale (a few tens of atomic layers), the motion of electrons in the growth direction is quantized in discrete energy states for electrons trapped in the potential well. In the direction parallel to the heterostructure layers the motion is not restricted and electrons behave like free electrons with an effective mass that is roughly the same they would have in the bulk well material. Electrons in this situation are termed bidimensional electrons, because their motion is restricted in two dimensions in the heterostructure layers.

When multiple quantum wells are grown next to each other, the situation is more complex and coupling between neighboring quantum wells can lead to the formation of narrow energy bands along the growth direction (minibands) allowing electrons to effectively move across the potential barriers. Electrons that possess an energy higher than the barrier behave like free three dimensional electrons in bulk, so they can travel across the heterostructure layers. Anyway the presence of the quantum wells affects their energy-momentum relation (dispersion relation), that can deviate substantially from that of a classical particle.

The ability to calculate the energy spectrum of semiconductor heterostructures is the first fundamental step that has to be taken in order to design quantum well detectors and other optoelectronic devices. Since the majority of infrared intersubband devices are unipolar and n-type, the system of interest is represented by the electrons in the conduction band only. This simplification allows us to use a one-band envelope function approximation [2] for the wavefunction of electrons in the conduction band only, where all the details of the underlying semiconductors (valence band and remote bands) are described by the effective mass and its dependency on position. Such an approximation is valid as long as the confinement energies are small compared to the energy gap of the materials involved. Within this limits the three-dimensional wavefunction of electrons $\Psi(z, \mathbf{r}_\parallel)$ can be decomposed in the product of a z-dependent function and a plane wave with planar wavevector $\mathbf{k}_\parallel$:

$$\Psi(z, \mathbf{r}_\parallel) = \psi(z) e^{-j\mathbf{k}_\parallel \cdot \mathbf{r}_\parallel} \tag{1.7}$$

where the z-dependent part $\psi(z)$ satisfies the 1D Schrödinger equation

$$\left(-\frac{\partial}{\partial z} \frac{\hbar^2}{2m^*(E, z)} \frac{\partial}{\partial z} + V_{cb}(z) + \phi(z) \right) \psi_\nu(z) = E_\nu \psi_\nu(z). \tag{1.8}$$

Here m^* is the effective mass, V_{cb} is the conduction band profile, $\phi(z)$ is the electrostatic potential due to the applied bias, and ν is a quantum number used to label eigenvalues. The two-dimensional energy dispersion can then be written as

$$E_\nu(\mathbf{k}_\parallel) = E_\nu + \frac{\hbar^2 |\mathbf{k}_\parallel|^2}{2m^*} \tag{1.9}$$

and is schematically plotted in the center panel of Fig. 1.1.

The main ingredients of Eq. (1.8) are the conduction band profile $V_{cb}(z)$ and the effective mass profile $m^*(z, E)$. The conduction band profile typically has the piecewise constant shape depicted in the left panel of Fig. 1.1, where the eigenstate energies are also shown. This ideal square shape is of course an approximation, as in a real device the actual shape of the interface between the two materials depends on the growth dynamics and thus can be different even for two nominally identical samples grown in different MBE machines. Anyway, since this deviations cannot be realistically taken into consideration without the introduction of empirical parameters, sharp interfaces are usually assumed. This is even more justified in THz

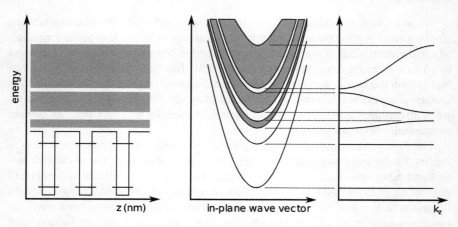

Fig. 1.1 Schematic of the conduction band profile with bound energies and continuum above the barriers (*left*). In-plane parabolic dispersion relation (*center*). Growth-direction dispersion relation (*right*) showing the quasi-3D nature of the continuum

devices where the barriers and wells are very thick, in the tens of nm range, making atomic-layer fluctuations negligible.

The dependency of the effective mass on the position and energy encodes all the information on the atomic structure of the underlying materials. The crudest approximation is to assume a constant mass, an approach that is justified in relatively large gap semiconductors (like GaAs) when the confinement energies are small and the wavefunctions do not significantly penetrate in the barriers. When dealing with smaller gap materials or when simulating quantum wells with high confinement energies, non-parabolicity effects become important and those can be included by considering the effect of valence bands as an energy-dependent effective mass. It can be shown that a three band model including light-hole and split-off band is equivalent to a one band model with the following effective mass energy dependence [4].

$$\frac{1}{m^*} = \frac{1}{m_0}\left(\frac{2}{3}\frac{E_P}{E - E_{\text{lh}}} + \frac{1}{3}\frac{E_P}{E - E_{\text{so}}}\right) \qquad (1.10)$$

where E_{lh} and E_{so} are the light hole and split-off band edge energies respectively, and E_P is the Kane energy, describing the strength of the coupling between conduction and valence band. More accurate descriptions can be obtained by including more electronic bands in the model, as described in detail in the book from Bastard [2]. In any case a model can only be as accurate as its input parameters at most, that is band alignments and effective masses (or equivalently the Kane parameters) which can be difficult to determine experimentally. In practical terms, it is typically not realistic to aim for a bound state energy accuracy of less than 1 meV. Since this corresponds to roughly 250 GHz, this error is particularly prominent when designing quantum well detectors in the few THz range.

The two-dimensional nature of subbands in quantum wells has a direct influence in the energy density of states of the system. It can be shown that unlike a bulk material where the density of states is proportional to $(E - E_c)^{3/2}$, in two-dimensional systems the density of states is constant for each subband $\rho = \frac{m^*}{\pi \hbar^2}$. Each successive subband adds the same amount of states to the spectrum, so that the final density of states can be written as

$$\rho(E) = \sum_{\nu} \frac{m^*(E_\nu)}{\pi \hbar^2} \theta(E - E_\nu) \tag{1.11}$$

where $\theta(E)$ is the unit step function and the general case of an energy-dependent effective mass has been considered.

1.2.2 Doping

Doping is a fundamental aspect of a quantum well detector design and operation. In III–V semiconductor systems n-doping is usually obtained with the inclusion of Si in the alloy. Once the Si donors are ionized the electrons will start populating the available states in the conduction band. The relation between the Fermi level and the number of electrons is particularly simple in quantum wells thanks to the piecewise constant density of states. When n_s is the sheet density of electrons in the well (the volume density multiplied by the doped layer thickness) the Fermi level in a single subband can be found as

$$E_F = \frac{\pi \hbar^2}{m^*} n_s \tag{1.12}$$

While in bulk systems n-doping only results in a shift of the Fermi level toward the conduction band, in quantum wells the situation can be more complex, depending on the doping location. The doping levels usually employed are such that electrons only populate the ground subband of quantum wells, up to a Fermi energy of a few meV at 0 K. This means that all electrons provided by the doping are localized in the wells. Depending on the location of the donors, two situations can be distinguished.

- If the donor impurities are included in the well material, then ionized impurities and electrons will coexist in the well. In this situation there is no physical charge separation but the electrons can interact with the ionized impurities and be scattered by them in their motion.
- On the other hand, if doping is included in the barriers and the donors are ionized, the electrons will in any case populate the quantum well states, being localized in a different place with respect to the fixed donors.

In the second case a layer of positively charged donors is left in the barriers, and a layer of negatively charged electrons forms in the well. This charge separation introduces an internal electrostatic field that has to be taken in consideration when calculating the conduction band profile (the $\phi(z)$ term in Eq. (1.8)). This technique is called modulation doping and is used to prevent scattering of electrons by the donor impurities, so to obtain higher electron mobility and lifetime.

1.3 Intersubband Transitions

An understanding of the interaction between electrons and light is fundamental in quantum well detector development. Intersubband transitions are qualitatively different from interband ones since the dispersion relation of the initial and final states are parallel, resulting in a single possible absorption frequency that depends on the thickness of the quantum well. This ability to tune the spectral properties by geometry rather than material composition is the most striking advantage of intersubband compared to interband devices. Again the subject is very broad and the interested reader can refer to the broad review by Ando, Fowler, and Stern [5].

1.3.1 Absorption Lineshape and Selection Rules

When the amplitude of an electromagnetic plane wave is small enough that its interaction with electrons can be treated perturbatively, the transition rate from state j to state i can be written in terms Fermi's golden rule

$$W_{ij} = \frac{2\pi}{\hbar^2} |\langle \psi_i | \frac{q}{m^*} \mathbf{A} \cdot \mathbf{p} | \psi_j \rangle|^2 \delta(E_{ij} - \hbar\omega) \qquad (1.13)$$

where $\mathbf{A}$ is the electromagnetic vector potential associated with the wave, $\mathbf{p} = -i\hbar\nabla$ is the momentum operator, and $E_{ij} = E_j - E_i$ is the energy separation between the initial and final states. In the so-called dipole approximation, the spatial variation of the electromagnetic plane wave is slow enough (with respect to the extension of the electron wavefunction) to be neglected and the vector potential is only a function of time characterized by the angular frequency ω.

When considering wavefunctions in the form Eq. (1.7) the matrix element in Eq. (1.13) is nonzero only if the electromagnetic wave has an electric field component in the confinement direction (the growth direction in quantum well systems). This is the well-known polarization selection rule for intersubband transitions, a key feature that dictates how intersubband optoelectronic devices must be designed. A relevant dimensionless quantity characterizing the strength of the transition is the so-called oscillator strength

$$f_{ij} = \frac{2}{m^* E_{ij}} |\langle \psi_i | p_z | \psi_j \rangle|^2 = \frac{2m^* E_{ij}}{\hbar^2} |\langle \psi_i | z | \psi_j \rangle|^2 \qquad (1.14)$$

in terms of which it is possible to compute the absorption coefficient α_{ij} which represents the fraction of electromagnetic energy that will be transferred from the electromagnetic wave to the electron population by exciting electrons from the state j to the state i:

$$\alpha_{ij} = \frac{q^2 h}{4\varepsilon_0 \varepsilon_r m^* c} \frac{\sin^2 \theta}{\cos \theta} f_{ij} n_{s,j} \frac{1}{\pi} \frac{\Gamma}{(E_{ij} - \hbar\omega)^2 + \Gamma^2}. \qquad (1.15)$$

Here $n_{s,j}$ is the sheet electron density in the lower subband, θ is the incidence angle with respect to the growth direction, ε_r is the refractive index of the quantum well medium. An characteristic feature of intersubband absorption is that it has a sharp spectrum centered around the energy separation between subbands. The origin of this feature lies in the fact that the two subbands involved in the photon absorption process are parallel in k-space, and thus the absorption happens at the same energy, regardless of the electron momentum, as depicted in the right panel of Fig. 1.2. It is also important to note that the absorption coefficient is linearly dependent on the electron density, making the quantum well doping level a key parameter to determine the optical characteristics of the transition.

In Eq. (1.15) the Dirac delta that was present in Eq. (1.13) has been replaced with a normalized Lorentzian with full width at half maximum Γ. This transition broadening is related to the finite lifetime of the states involved in the transition. In an ideal system with no interactions such broadening would be zero, but in a realistic quantum well electrons interact with lattice vibrations and with other electrons. A summary of the strongest interactions is given in Fig. 1.2. Electron-longitudinal opti-

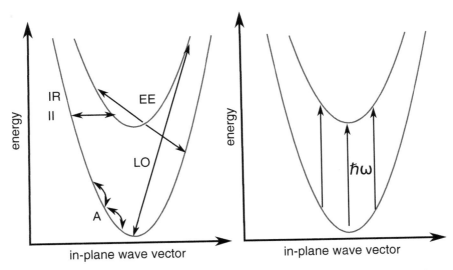

Fig. 1.2 *Left* Various scattering mechanisms that affect electron motion in subbands. *Right* photon absorption between subbands happen at the same energy, regardless of the electron wavevector

cal (LO) phonon interaction is an inelastic process where an electron exchanges a quantum of mechanical oscillation (a phonon) with the crystal lattice. Typical energies of phonons in III–V semiconductors are around 30–40 meV and their dispersion is flat enough for them to be considered as waves at a well-defined frequency and arbitrarily big wavevector. It has to be noted that optical phonon energies are higher than THz transitions energies, and thus the transition is always diagonal in k-space as depicted in the left panel of Fig. 1.2. Since the strength of electron–phonon interaction becomes smaller for larger exchanged wavevector, the result is that for THz intersubband devices electron phonon interaction is not a dominant scattering mechanism, as it is for mid-infrared devices. Acoustic phonons (A), on the other side, have the linear dispersion typical of acoustic waves and thus are available at all energies. For this reason they are very effective in making electrons dissipate energy within the subband, making them thermalize towards the bottom of the band. In addition to phonons, deviations from the ideal square potential landscape that arise because ionized donor impurities (II) and interface roughness (IR) also interact with electrons, introducing elastic scattering processes between subbands. These do not cause electrons to dissipate energy, since they do not involve emission or absorption of vibrations, but can still scatter electrons horizontally in k-space causing transitions between subbands.

Finally when the doping density is high enough that electrons start to feel each other's electric field, electron–electron (EE) scattering becomes important. This process involves two electrons and can provide both energy and momentum transfer. Unuma et al. [6] carried out a systematic investigation of these effects on the linewidth. They found that interface roughness is the dominating factor in mid-infrared transitions, because it broadens the width of the excited state by a large amount. In THz quantum well detectors the effect of roughness is smaller because the wells are bigger with respect to mid-infrared devices, leading to a proportionally lower influence of roughness. If the intersubband transition energy is below the optical phonon, then also optical phonon scattering is almost absent. Acoustic phonons and disorder effects like the impurity potential are supposed to be dominating in THz-detectors.

1.3.2 Many-Body Effects

In the previous discussion the electron population has been considered as an ensemble of non-interacting particles. This assumption allows one to write down a so-called single particle Hamiltonian and to easily find the eigenstates of the system by solving the Schrödinger equation in one variable. In real systems, especially when the doping is high, Coulomb interactions between electrons play a substantial role, to the point that the system cannot be anymore considered as formed by non-interacting particles. Two main effects arise in these conditions. First of all electrons feel the electric field of other electrons as an additional potential. When the electron density is large enough a good approximation is to assume that every electron feels an average electric field given by the charge distribution due to all other electrons. This introduces an additional potential in the $\phi(z)$ term in Eq. (1.8) that depends on the wavefunction and satisfies Poisson's equation. The eigenstates of the system have to be found by solving the following system of differential equations selfconsistently:

$$\left(-\frac{\partial}{\partial z}\frac{\hbar^2}{2m^*(E,z)}\frac{\partial}{\partial z} + V_{cb}(z) + \phi(z)\right)\psi_\nu(z) = E_\nu\psi_\nu(z) \tag{1.16}$$

$$\phi(z) = \frac{d^2}{dz^2}\frac{q\sum_\nu |\psi_\nu(z)|^2}{\varepsilon_0\varepsilon_r} \tag{1.17}$$

In addition to this electrostatic interaction electrons are correlated by the so-called exchange interaction. This effect is related to the parity of the many-body wavefunction and has no classical analogue. The effect can also be included as an additional potential dependent on electron density, in the so-called local density approximation. Bandara et al. [7] derived an approximate expression for a single QW with infinitely high potentials. In practical terms the corrections are typically of the order of a few meV for in-well doping, and can be a few tens of meV for modulation doped structures. These deviations are thus important only when there is significant donor-charge separation, or when the transition energies are very small, like in THz quantum well detectors.

The many body effects discussed so far affect the value of the eigenstate energies, with the effect of shifting the transition energies if the energy renormalization differs for every state, as is usually the case. Anyway intersubband absorption in itself is a collective process of the electron gas that cannot be accurately described only in terms of an isolated electron absorbing a photon and jumping to an excited state. The photon absorption process for 2D electron gases (2DEG) is more complex as the interaction of each electron with the electromagnetic field is screened by the presence of the other electrons. As a result the observed optical absorption frequency depends on the electron density. A common formula for the calculation of the shifted transition $\tilde{E}_{ij}$ energy is

$$\tilde{E}_{ij}^2 = E_{ij}^2(1 + \alpha - \beta) \tag{1.18}$$

where α accounts for the interaction of electrons with the other electrons oscillating in the applied electric field (depolarization shift) and β accounts for a similar interaction with the "hole" left behind by the excited electron (exciton shift). In reality, there are infinite correction terms with alternating sign and decreasing magnitude, as explained in [5], but those are not usually taken into account. Expressions for α and β can be found in the book by Helm [8] where he also reports a calculated example of a well-doped 8-nm GaAs well with $Al_{0.3}Ga_{0.7}As$ barriers and 2×10^{18} cm^3 doping over 5 nm. The ideal square well calculation produces a 111.2 meV ground to first excited state energy. The self-consistent Schrödinger-Poisson solution changes it to 112.3 meV, and including exchange-correlation effects results in 114.6 meV. The depolarization effect further pushes the absorption to 123.4 meV, and the exciton shift reduces it to 120.3 meV. The global effect of all interactions is to move the resonance to a larger energy, consistent with experiments, but it can be noted how the depolarization shift gives a major correction that anyway is an overestimation.

1.4 Transport in Multi-Quantum-Well Systems

The problem of simulating carrier transport in semiconductor quantum well systems is still an open and very active research field due to the variety of possible configurations and the number of interactions between the electron population and the environment. Quantum well photodetectors are simpler than most other semiconductor structures, but the description of carrier transport among them is far from being simple. In fact, the presence of the quantum wells requires a model to take into account trapping and detrapping of carriers, but since the conduction happens in the continuum across the barrier, a strong drift-diffusion component is present. In the following, three models that have been developed for the task are reported.

1.4.1 Emission-Capture Model

In their famous book on quantum well detectors, H. C. Liu and H. Schneider [1] present a very effective model for the evaluation of detector parameters based on the so-called emission-capture model. The building block of the model is the quantum well and its neighboring barriers, forming one period of the multi-quantum-well structure.

The model is based on a few fundamental assumptions: (a) the interwell tunnelling contributes negligibly to the dark current, (b) the electron density in each well remains constant, (c) the heavily doped emitter serves as a perfectly injecting contact, and (d) mainly one bound state is confined in the quantum well. These assumptions are all satisfied in common QWIP geometries at low bias voltages.

In a photoconductive QWIP the current is due to the flow of electrons above the barriers, that behave as free electrons in a three-dimensional space. As electrons come close to the quantum wells, they can loose energy by scattering and be trapped in the well. Similarly, a trapped electron can acquire sufficient energy to escape the well by interacting with the lattice and start drifting towards the contacts. Figure 1.3 schematically depicts the situation. In addition to the large drift current j_{3D}, small emission (j_e), and capture (j_c) currents can be introduced in the vicinity of the well, which represent carriers that instead of flowing through the quantum well are trapped inside and then emitted again. In stationary conditions, when the population of the bound state is constant, the condition $j_e = j_c$ holds. In addition, since the current is the same throughout the structure, j_{3D} and j_e (or j_c) are related. A trapping or capture probability p_c can be defined for an electron traversing a well with energy larger than the barrier height which allows one to write the relation

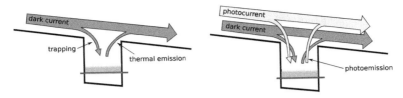

Fig. 1.3 Schematic of the emission-capture model from ref. [1]. In addition to the flow of unbound electrons above the quantum wells, capture and emission currents are present due to the interaction of electrons with the crystal lattice. The presence of light causes additional trapping and emission that modulate the drift current in the three-dimensional continuum

$$j_{3D} = j_c + (1 - p_c)j_{3D} = j_e + (1 - p_c)j_{3D} \tag{1.19}$$

This is the key relation of the model since it allows to calculate the total current through the device once the emission current and the capture probability are known. In fact from the previous equation it follows that

$$j_{3D} = \frac{j_e}{p_c} \tag{1.20}$$

The key point is then to evaluate the emission current and the capture probability. The latter is written in terms of characteristic times of the system, by introducing a capture time τ_c that describes the average time it takes for an electron around the well to capture, and the scattering time τ_s that describes the average time between two scattering events that can push a bound electron out of the well. With these assumptions it is possible to obtain an expression for the dark-current density

$$j_{dark} = q \frac{n_s}{L_p} \frac{\tau_c}{\tau_s} v(F) \tag{1.21}$$

where L_p is the sum of well and barrier thickness (one period of the multi-quantum well structure) and $v(F)$ is the velocity of electrons above the barriers as a function of the applied electric field, that can be evaluated once the mobility of electrons is known. The model allows for a very natural inclusion of electron–photon interaction by means of an additional capture-emission current due to absorbed photons (second panel of Fig. 1.3), leading to an expression for the specific detectivity in terms of the characteristic times of the device

$$D^* = \frac{\lambda}{2hc} \frac{\eta}{\sqrt{N}} \sqrt{\frac{\tau_s}{n_s}}. \tag{1.22}$$

Here N is the number of quantum wells and η is the absorption quantum efficiency.

The advantage of this model lies in the fact that besides providing good fits with experimental data, it also provides useful insights in the important physical processes governing quantum well detectors.

1.4.2 Kinetic Model

A kinetic model based on the three-dimensional Boltzmann transport equation for THz QWIPs was proposed by Castellano et al. [9]. Contrary to the emission-capture model presented before, in the kinetic model there is no distinction between the two-dimensional quantum well states and the three-dimensional continuum above the barrier. A full three-dimensional band structure, or dispersion relation of electrons, is obtained by solving the Schrödinger equation in reciprocal space assuming the potential is infinitely periodic, which is a good approximation when the number of quantum wells is high. The band structure $E_\nu(k_\parallel, k_z)$ of the system is the one sketched in Fig. 1.1. The center panel depicts the in-plane parabolic dispersion relation, whose analytical form is given in Eq. 1.9. The right panel depicts the dispersion relation along the growth axis z, where minibands are formed due to the periodicity of the underlying potential. In this picture, each state labeled by the subband index ν and wavevector $\mathbf{k} = (k_\parallel, k_z)$ can be populated by electrons. The temporal evolution of the electron population distribution $f_\nu(k_\parallel, k_z)$ obeys the Boltzmann transport equation

$$\frac{\partial f_\nu(\mathbf{k})}{\partial t} = \frac{e}{\hbar}\mathbf{F} \cdot \nabla f_\nu(\mathbf{k}) + \sum_{\nu'=1}^{N} \int \left[P_{\nu\nu'}(\mathbf{k}, \mathbf{k}')f_{\nu'}(\mathbf{k}') - P_{\nu'\nu}(\mathbf{k}', \mathbf{k})f_\nu(\mathbf{k}) \right] d\mathbf{k}' \quad (1.23)$$

The $P_{\nu\nu'}(\mathbf{k}, \mathbf{k}')$ is the probability per unit time that an electron initially on subband ν' at wavevector $\mathbf{k}'$ undergoes a transition to subband ν at wavevector $\mathbf{k}$. Average values of physical quantities, like the current, can be calculated from the distribution function obtained by solving Eq. (1.23) in stationary conditions. The electron drift is provided by the electric field $\mathbf{F}$, that acts on the distribution function by translating it in k-space in the direction of the applied field, thus giving a nonzero average momentum to electrons which translates to a current.

Absorption of light is included at the microscopic level by introducing transition probabilities between the various k-space states according to the spectral power density of the incident radiation. This allows to microscopically calculate the current induced by the optical background at 300 K and evaluate the T_{BLIP} of the device.

The scattering probabilities deriving from the interaction with the environment can be calculated from the scattering mechanisms depicted in Fig. 1.2, but in this case a simplified scattering mechanism is considered. Instead of calculating the microscopic transition probabilities due to all possible scattering mechanisms, a procedure that needs a lot of material parameters that have to be determined experimentally, a single effective scattering process is assumed. The functional form of the scattering probabilities is chosen in order to ensure proper thermalization of the electron system in absence of external bias or optical field. The only free parameter of the model is the characteristic time that sets the strength of the effective scattering. Such time can be obtained by fitting the T_{BLIP} of a real device similar to the one that is to be simulated.

The simplified scattering mechanism implemented sets one of the limitations of the model. In fact, considering scattering characterized by a single time is a good approximation only when there are no energy-dependent processes with sharp thresholds like longitudinal optical phonon emission. This is not a problem in the simulation of THz QWIPs, where such processes are of minor importance. In any case such processes can be easily included by plugging the appropriate scattering matrices into the Boltzmann equation. The strength of this model lies in its ability to simulate realistic experiments with arbitrary optical power spectral densities (for example a laser field) and structures with arbitrary band structure besides being also suitable for time-dependent simulations.

1.4.3 Tunneling Currents

Up to this point the attention has been focused on the photocurrent and the drift dark current caused by the thermally excited electrons, while interwell tunnelling currents have always been neglected.

For devices operating in the liquid nitrogen temperature range the dark current is caused by thermally excited carriers and thus shows an exponential dependence on temperature. The most common way of reducing the dark current in present-day QWIP-based infrared imagery systems is thus to reduce the operating temperature in the liquid nitrogen range until the desired performance requirements are met.

In some aerospace and defense applications temperature is reduced further into a regime where the dark current is dominated by the tunnelling rather than thermionic emission. Nevertheless because such applications usually involve very low photon fluxes, the reduction of dark current remains a crucial issue that cannot be addressed by further temperature reduction since the tunnelling currents are independent of it.

Recent investigations on the subject [10, 11] were able to reproduce the dark current-voltage characteristics of the same mid-infrared QWIP including tunnelling currents using different approaches.

In [10] the problem is treated as an extremely weakly coupled superlattice and the current is computed by assuming the ground states of every quantum well couple to form a very narrow miniband. At higher voltages direct tunnelling into the continuum becomes dominant and is accounted for by transmission through multiple triangular barriers formed by neighboring quantum wells. Figure 1.4 shows the comparison between experimental and simulated IV curves for a mid-infrared QWIP for different temperatures.

In [11] the tunnelling current for the same device is calculated by assuming scattering driven hopping transport between successive ground states, due to interaction of electrons with phonons, impurities, and roughness. The approach is completely

Fig. 1.4 Simulated (*solid line*) and measured (*dotted line*) current voltage characteristic of a mid-infrared QWIP for different temperatures [10]. Tunneling currents appear as a low voltage plateau at low temperatures

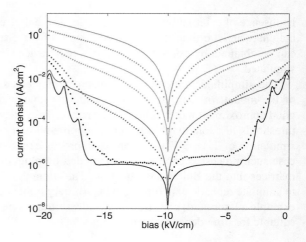

different from the one adopted in [10] but the agreement with experimental data is good. A straightforward way to reduce the tunnelling currents would be to increase the barrier thickness; however, doing so would require a thicker structure in order to maintain the same number of periods and the same absorption. Authors in [12] showed how the tunnelling current depends on impurity scattering and that it can be reduced by appropriate engineering of the doping location.

1.5 Quantum Well Detectors

A quantum well infrared detector (QWIP) is obtained by combining the ingredients presented in the previous sections into a device. A stack of quantum wells separated by barriers is grown by MBE typically on a GaAs substrate. A metallic contact on top and a buried doped layer on the bottom of the growth provide the electrical contact and injection of carriers. Once the device is exposed to light, absorption of photons produces a photocurrent. In the following, the QWIP design process as explained in [1] is presented, and then the current state of the art is reviewed.

1.5.1 Optimum design

Designing an optimum photoconductive QWIP involves choosing the well width, barrier height, barrier width, well-doping density, and number of wells. The simplest structure is the one based on GaAs/AlGaAs square quantum wells. The well region is GaAs, and the barrier is $Al_xGa_{1-x}As$, the height of the barrier is controlled by the Al fraction x. It can be shown that the optimum well shape is the one having the first excited state in resonance with the top of the barrier. The well width and barrier height (Al fraction) are fixed by this rule and by the desired detection wavelength. This process requires the ability to calculate the energy states of the quantum wells, with all the implications explained in Sect. 1.2.1.

As for the well doping, the design rule is that to maximize the detector-limited detectivity, the doping density should be such that the Fermi energy is $E_f = 2k_BT$, where T is the desired operating temperature. To maximize the blip temperature instead the requirement is $E_f = k_BT$.

The barrier should be thick enough to ensure that the interwell tunnelling current is completely negligible in comparison with the background photocurrent. The tunnelling currents exhibit a decreasing exponential dependence on the barrier thickness, but a precise evaluation of tunnelling currents requires appropriate modeling, as explained in Sect. 1.4.3. However, practical MBE growth concerns require the total detector thickness to be not too big, of the order of a few microns at most. This sets an upper bound on the barrier thickness and the number of quantum wells that can be grown.

By analyzing Eq. (1.22) one can see how the detectivity increases with increasing number of quantum wells. This is because the absorption quantum efficiency η is proportional to the number of wells, saturating to unity as N becomes very large. Anyway, the $\sqrt{N}$ factor at the denominator will cause a reduction of detectivity if the absorption is saturated. There is then an optimum that has to be evaluated for each design. As mentioned before, the number of wells is also limited by practical fabrication arguments that come into place in this phase.

1.5.2 Optical Coupling

Up to now only the electrical characteristics of quantum well detectors have been discussed, without devoting attention to the optical coupling, which is indeed a key aspect of photodetector development.

The main concept to keep in mind with intersubband optoelectronic devices is that absorption is p-polarized: the electric field must have a component in the growth direction, perpendicular to the heterostructure layers. This means that, unfortunately, it is not possible to absorb light normally incident on the QWIPs surface, because the electric field in this case lies parallel to the detector surface. In addition, since the bias current also needs to flow perpendicular to the layers, an electrical (metallic) contact on the top of the device and another one on the bottom are needed, and these would reflect away the incoming radiation.

There are two common solutions to this problem, that are used for different applications. In the simplest case, suitable for single pixel detectors or during device design and in general in intersubband transitions research, the semiconductor substrate on which the quantum wells are grown is used as a prism or waveguide based on total internal reflection, as depicted in Fig. 1.5. The substrate is polished at 45° and light is focused on the polished facet perpendicular to it. The refractive index

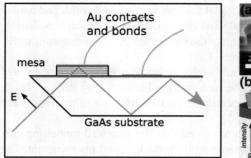
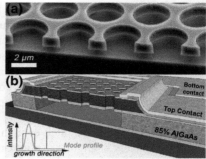

Fig. 1.5 (*left*) The substrate coupling geometry, used mainly in research as it allows to quickly test a device. (*right*) Photonic crystal couplers [15] produce strongly enhanced responsivity at discrete frequencies

of substrates is typically around 3.5, giving 60 % transmission at the interface. The light that is transmitted through the substrate reaches the quantum well region, where it is reflected back by total internal reflection or by the top metallic contact. In this configuration the light has a component in the growth direction so photons can be absorbed. The bottom contact of the QWIPs is a 100 nm thick buried n-doped layer that is contacted from the side and is partially transparent to radiation. The QWIP is usually processed into a mesa shape by lithography and etching when a single pixel detector is needed, otherwise the quantum well layer can be left untouched in order to allow for multiple interactions with light and increase the absorption. This is a commonly used technique in intersubband absorption measurements.

In commercial imaging applications another solution is adopted. A diffraction grating is produced in close proximity to the quantum wells, to diffract the light laterally, introducing electric field components in the growth direction. Diffraction gratings are frequency selective, so appropriate designs are needed in order to match the bandwidth of the detector and the grating. This solution allows absorption of normal incident light and at the same time the development of large area focal plane arrays. Gratings can be either etched in an additionally grown layer after the quantum wells [13] or made of metal [14].

More exotic solutions are being explored to enhance the response of QWIPs. A notable one is the recent demonstration of photonic-crystal couplers [15], shown in the right panel of Fig. 1.5. The concept is to create a very high quality factor cavity by building a photonic crystal structure suspended in air. The extremely high mode confinement traps the photons in the cavity and lets them interact with the QWIP for a longer time, enhancing the absorption. A marked increase in responsivity was observed even at very low doping levels. Of course this happens just at the resonance frequency of the photonic crystal, where a laterally localized mode is present.

1.5.3 State of the Art

QWIPs are at the moment a commercial technology for the mid-infrared spectral band, with large megapixel focal plane arrays available with very good performance [1, 16]. On the other side, THz QWIPs are much less developed, having been only recently demonstrated for the first time.

The first demonstration of QWIPs below the optical phonon frequency was from the group of H. C. Liu [17, 18]. Background limited detection was obtained with three QWIPs with different detection wavelengths in the terahertz region. The demonstration of THz QWIPs [17] failed to achieve BLIP detection even at liquid helium temperature due to the large amount of dark current in the device, which was attributed mainly to interwell tunnelling. In THz QWIPs the barriers are very low since the confinement energies must be in few tens of meV range. This means that the Al content in barriers is of the order of 1–6 %. A very low barrier then becomes more transparent to electron tunnelling, leading to big tunnelling currents. In [17] the problem was solved by employing very thick barriers (50–90 nm) with the effect of the tunnelling current dropping below the background current. Blip temperatures of 17, 13, and 12 K were achieved for peak detection wavelengths of 9.3, 6.0, 3.3 THz, respectively. Such very low blip temperatures are also due to the very small electron confinement energy in the well, that causes thermal dark current to quickly dominate as the temperature raises.

Since then research has proceeded in the direction of trying to clarify the mechanisms that govern the THz QWIP operation and to improve performance. It was proposed [19] that employing quantum wells with more than one confined state would increase the T_{BLIP} due to a reduction of dark current. This claim has not yet been verified but two-level THz QWIPs have been explored [20] observing enhanced optical nonlinearity leading to the demonstration of the device as a quadratic autocorrelator for far-infrared picosecond pulses at around 7 THz.

The authors in [21] observed a strong enhancement of the THz response in a GaAs/AlGaAs quantum well photodetector under magnetic fields applied in the growth direction. In this conditions, Landau quantization of the in-plane electron motion in the quantum wells is the predominant underlying mechanism for the improvement of photoconductive gain. Other minor contributions come from an increase in the detector differential resistance.

Another common problem of THz QWIPs is the lack of absorption, due to the fact that the Fermi level cannot be matched to the optimal value of $k_B T$ because that would exceed the confinement energy. However, even at lower doping levels many-body effects start to play a crucial role as shown in [22] where band structure calculations on heavily doped THz QWIPs show that the second subband falls into the quantum well with an increase of the Si doping concentration. The observed effect is a blueshift and broadening of the photocurrent peak. The importance of many body effects has also been reported in [23] where authors found a large discrepancy between the theoretical and experimental photoresponse peak positions when many-body interactions are not taken into account. The calculated results agree with the experimental data quantitatively when exchange-correlation and depolarization effects are added within the local density approximation.

The use of alternative materials is also being explored. Patrashin et al. [24] designed an InAs/Ga$_{0.6}$In$_{0.4}$Sb superlattice (SL) material for terahertz-range photodetectors. Depending on the thicknesses of the InAs and Ga$_{0.6}$In$_{0.4}$Sb layers, the SL energy gap can be adjusted to be between 8 and 25 meV, which corresponds to a cut-off frequency from 2 to 6 THz. Such material poses many challenges both from the design and growth point of views, due to the presence of strain and the type-II band alignment which bring the valence band into play.

Recently GaN is also drawing attention in the THz world, mainly because of its high energy longitudinal optical phonon the would allow operation at higher THz frequencies. Anyway the material technology is still not mature enough to provide devices, but recently THz absorption has been measured in GaN/AlGaN quantum wells [25]. Since GaN-based materials have a very strong material polarization, heterostructure have very big internal electric fields. In [25] structure has been optimized to approach a flat-band potential in the wells to allow for an intersubband absorption in the terahertz frequency range and to maximize the optical dipole moments.

1.5.4 Quantum Cascade Detectors

The most common type of QWIP is a photoconductor, as showed in the preceding sections. Anyway as explained in Sect. 1.1 there are intrinsic advantages in photovoltaic detectors that are not present in photoconductors. These are the improved noise figure of merit and the potential absence of any dark current. In order to have a photovoltaic QWIP an asymmetry in the potential needs to be introduced, such that photoexcited electrons spontaneously diffuse toward one of the contacts. This is achieved in quantum cascade detectors [26] by building a path for the electrons through a series of coupled quantum wells.

As the confinement energy becomes smaller at longer wavelengths, the dark current becomes higher because the average thermal excitation energy of electrons allows them to easily escape the well. In QWIPs this cannot be avoided since the design rule is to have the excited state of the well in resonance with the top of the barrier, thus fixing the depth of the ground state.

In quantum cascade detectors this restriction is removed and the quantum well ground state is allowed to be deeper in the well, thus reducing the amount of thermally excited electrons. In a normal QWIP design this would result in a reduced photocurrent since the photoexcited electron would not easily escape the well. In a quantum cascade detector multiple scattering events take place sequentially between adjacent quantum wells until the electron is injected back in the ground state of the next period. This process, schematically depicted in Fig. 1.6, repeats for every period in the structure, with the electron absorbing a photon and moving to the next period, until the collector contact is reached.

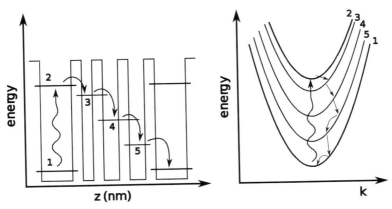

Fig. 1.6 Schematic of a quantum cascade detector. Absorption from subband 1 to 2 is followed by scattering through state 3 ,4, 5 until the next ground state is reached. On the *right*, the equivalent process is in k-space. The photon absorption is vertical, while energy loss via scattering is accompanied by momentum randomization

The device is a photovoltaic one since no external bias is in principle needed for the current to flow: the energy for electron motion is given by the absorbed photons. The immediately apparent drawback of this concept is that the responsivity is reduced proportionally to the number of periods, since many absorbed photons are needed for an electron to reach the contacts. On the other hand, being a photovoltaic detector, it benefits from the intrinsically better noise performance, and also a reduced dark current.

In practice, a small bias voltage usually helps to prevent backflow of electrons by lowering the energy of the ground states of successive periods, improving performance.

One of the key aspects of QCD performance is the ability to efficiently remove a photoexcited carrier from the well and transfer it to the next period. The typical solution is to use resonant phonon extraction by setting the energy difference between the extractor equal to that of an optical phonon.

Quantum cascade detectors are being actively developed in all infrared spectral ranges. On the short-wavelength side, difficulties in the design of QCD arise because of the small conduction band discontinuity available in commonly used III-V ternary alloys, typically around 500 meV. A 2.14 μm detector was demonstrated [27], exhibiting a peak responsivity of the first one is 2.57 mA/W at 300 K. An alternative material system that is being explored for short-wavelength intersubband optoelectronics is the GaN/AlGaN system, which offers a big (1.9 eV) conduction band discontinuity [28]. The shortest-wavelength QCD demonstrated so far [29] is based on such material system, performing room temperature with a responsivity of 10 mA/W at 1.7 μm. Quantum cascade detectors in the mid-infrared, around 4 μm have reached noise performance levels comparable to those of QWIPs [30] with T_{BLIP} values around 120–150 K. Wide spectral range detection was also demonstrated in the mid-infrared with an InP-based device presenting a broad response between 4.7 and 7.4 μm [31]. At longer wavelengths, around 9 μm both AlGaAs based [32] and InP-based [33] devices have been developed. The operating temperatures in this region are around 50–70 K.

The long wavelength regime is where QCDs are most promising, as they provide a lower dark current with respect to photoconductive QWIPs. This allows for very long integration times in photon-starved applications like astronomic imaging. An InP-based QCD at 16.5 μm has been proposed in 2007 by Giorgetta et al.[34]. This wavelength corresponds to a transition of 75 meV, which allows only 2 transitions through optical phonons for depopulation. In a classical design this would imply a very short period with just two wells, with the result of having the ground states of successive periods coupled. In [34] the problem was solved by introducing a superlattice extractor that allows both efficient diffusion of photoexcited electrons and isolation between the periods. The maximum reported operating temperature is 90 K.

Concerning the GaAs/AlGaAs material system, a 14.3 μm device was demonstrated by Buffaz et al. [35]. Electron relaxation to the ground state in this case is provided by two coupled wells whose doublets are separated by the energy of a LO-phonon. The maximum detectivity was found at a temperature of 25 K and a bias of −0.6 V. Noise performances in this wavelength range are of the same order as QWIPs [36].

So far the only published THz quantum cascade detector is the one from Graf et al. [37], where they exploited quantum cascade lasers structures as THz detectors. The device operates at 84 μm with a responsivity of 8.6 mA/W and a detectivity of 5×10^7 Jones at 10 K. Recently, the interest in THz QCD has revived with works aimed at identifying the dominant transport mechanisms. In fact, since LO-phonon-driven scattering is absent, transport has to rely on elastic scattering between subbands and intraband relaxation. Lhuillier et al. [38] identified interface roughness as the dominant contribution to the current by simulating carrier transport by means of a hopping model, with interface roughness parameters extracted from TEM images.

1.6 Quantum Dot Detectors

Starting from a 3D bulk system, adding quantum confinement in one direction (layered heterostructure) produces a 2D system, a quantum well as explained before. If the motion of electrons is also contstrained in the plane of the layers, fully confined states are possible, where the electrons cannot propagate in any direction. Such systems are termed quantum dot, or zero-dimensional potential wells.

1.6.1 Quantum Dot Fabrication

The most widespread semiconductor quantum dots are the so-called Stransky-Krastanov (SK) self-assembled quantum dots [39, 40]. SK dots are produced by molecular beam epitaxy, very much like quantum wells, with the important difference that semiconductors of different lattice constant are grown on top of each other. In the SK process a layer of InAs starts to grow on GaAs, the lattice mismatch between the two materials being 6.6 %. Under such a big strain the growth of InAs cannot continue indefinitely, and as soon as the layer reaches a critical thickness it relaxes forming nm sized crystalline InAs droplets on the surface of GaAs [41]. The droplets are small enough for the electrons inside to be quantum confined in all three directions, exhibiting a purely discrete energy spectrum similar to that of an atom. The analogy with atoms is so strong that the terminology for the energy state labeling is actually mutuated by atomic orbitals, with the 2 spin-degenerate low lying states called *s states* and the first excited states termed p_x, p_y, and p_z.

The shape of SK dots is lens-like, depends on the particular growth condition, and shows large variations across the sample. This big size variability among different dots in the same layer is the main drawback of self-assembled quantum dots, resulting in a system with a large (around 40–50 meV) Gaussian distribution of the ground state energy and the s-p energy separation.

The energy density of states of a quantum dot ensemble is in principle the one of an atom ensemble. In practice, due to the above mentioned inhomogeneity the deltas are replaced by Gaussians that account for the size distribution of the dots and the corresponding variability in bound state energies. In addition, asymmetries in the dot shape will lift the degeneracy of p states, with the p_x and p_y states usually much lower than the p_z state, due to the much bigger lateral size of the quantum dot with respect to its thickness.

A high degree of uniformity can be obtained through substrate prepatterning prior to dot growth [42]. With this technique quantum dot nucleation is driven by artificial defects in the substrate introduced by lithography and etching.

Figure 1.7 summarizes the two growth techniques. (a) Self assembled quantum dots growth starts by a strained InAs layer on GaAs substrate. (b) When the critical thickness is reached, the InAs layer relaxes into droplets. (c) The process can be repeated again after GaAs regrowth. (d) Patterned dots growth starts by an etched substrate. (e) InAs grows faster in the etched pits than elsewhere. If the grown thickness of InAs is just below the critical thickness, dots are only formed in the substrate recesses.

The advantage of patterned dots growth lies in the fact that the dot characteristics (size, position, and confinement energy) can be controlled by lithography, leading to a very uniform array with inhomogeneous broadening energy down to 1 meV [43].

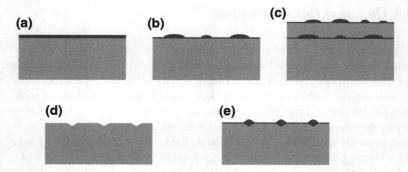

Fig. 1.7 Self assembled (**a–c**) versus patterned (**d, e**) quantum dots fabrication

1.6.2 Transitions in Quantum Dots

One of the most prominent differences between intersubband transitions in quantum wells and interlevel transitions in quantum dots is the lack of polarization selection rule in the latter. In fact, the polarization selection rules arise because of the need to conserve in-plane momentum during the transition, because of the translational invariance of the quantum well structure. In quantum dots such restriction is released because the potential is not translationally symmetric anymore. Quantum dots can thus in principle absorb light in any polarization. In practice, in self-assembled quantum dots the transition still shows some polarization selectivity. This is due to the typical lens shape of SK dots, with base diameter of 15–20 nm and 3–7 nm height. This causes the lateral confinement be much weaker than the growth direction confinement, making the dot resemble more a laterally small quantum well and introducing polarization selectivity.

Recently, Schao et al. [44] reported an enhancement of the s-to-p polarization ratio up to 50%. The result was made possible through engineering of the dot geometry via post-growth capping of the QDs with an InAlGaAs layer. The resulting shape of the dots is cone-like with a base diameter of 12 nm and an increased height of 8 nm.

1.6.3 Quantum Dot Detectors

Quantum dot infrared photodetectors (QDIPs) are conceptually similar to QWIPs, the only difference being that the quantum well is replaced by a layer of quantum dots, and optical transitions between zero-dimensional states provide the detection mechanism [45–47]. Despite their similarity, QDIPs are promising candidates to solve the problems that generally affect QWIPs. First of all the zero-dimensional confinement removes the polarization selection rule in absorption, leading to a detector that is insensitive to the polarization. In addition, the peaked density of states translates into

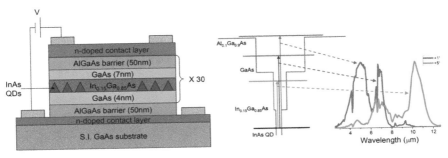

Fig. 1.8 Typical multi-period DWELL structure. Reproduced with permission from [49]

10–100 times longer carrier lifetimes with respect to QWIPs. This effect is due to the fact that the energy separation between QD states is generally bigger than the optical phonon energy, meaning that an electron in an excited state cannot scatter into the ground state by emitting a phonon. On the contrary, in a 2D system an electron can always emit a phonon if the subband separation is bigger than the phonon energy. This lifetime enhancement effect in quantum dots goes under the name of phonon bottleneck [48] and has been shown to be related to polaron formation due strength of electron phonon-interaction.

The most promising architecture for QDIPs is the so-called dot in a well (DWELL) detector, whose typical structure is shown in Fig. 1.8 from [49]. The InAs quantum dot layer (represented as triangles) is embedded in a GaAs quantum well, and the different periods are separated by AlGaAs barriers. The role of the barriers is to limit the dark current, while the presence of the GaAs well enhances electron capture and provides final states for optical transitions to occur. Since the formation of self-assembled quantum dots is driven by the strain between the InAs and GaAs layers, strain engineering techniques [50] are applied in order to allow the growth of multiple quantum dot layers without introducing crystal defects. The absorption coefficient of a quantum dot plane tends to be lower than that of a quantum well because of the small filling factor of quantum dots, and although devices with a single quantum dot layer are possible [51], repetition of the structure is beneficial because it provides higher absorption.

Due to the presence of the InGaAs wetting layer and the big lateral size of the SK dots, the bandstructure can be rather complex and multiple absorption peaks appear, corresponding to transitions between the quantum dot ground state and one of the quantum well states. Quantum dots in a well (DWELL) design [52] allows better control of the peak operating wavelength because the transition frequencies can be determined by the quantum well thickness, and also reduces the dark current [53, 54] thanks to the AlGaAs barriers.

The relative performances of QWIPs and QDIPs are summarized in Fig. 1.9, reproduced from [49]. In the left panel it can be seen how QDIPs are starting to surpass the performances of QWIPs in terms of operating temperature while right panel shows published detectivity data (circles are QWIPs, other symbols are QDIPs). The

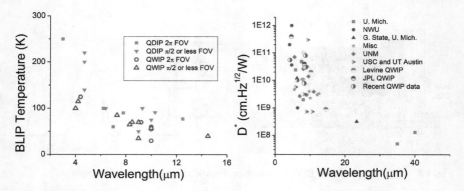

Fig. 1.9 QWIP and QDIP performance comparison. Reproduced with permission from [49]

process for the fabrication of QDIP focal plane arrays is the same as for QWIPs, the only difference being in the growth phase. Thanks to this technological compatibility, QDIP focal plane arrays in the mid-infrared operating up to 200 K have been demonstrated [55].

From the modeling point of view, QDIPs present additional challenges with respect to QWIPs. The bandstructure of quantum dots is intrinsically harder to determine because a fully 3D solution of the Schrödinger equation is needed, in addition the actual shape of the quantum dot is not precisely known and shows large variations across the sample. The tight confinement of electrons and the fact that just a few of them populate the dots make many-body effects much more prominent, further complicating the eigenstate calculation. Finally the transport properties depend on the complex interaction between 3D states above the barriers, two-dimensional wetting layer states and zero-dimensional quantum dot states. Lim et al. [56] have modeled QDIP transport by means of a diffusion-recombination formalism. Their findings support the empirical observation that electron capture times are longer that the transit time across one period above 70 K, translating into a photoconductive gain higher than 1. A more microscopic model was developed in [57] where the carrier wave functions and energy levels were evaluated using the strain dependent eight-band $\mathbf{k} \cdot \mathbf{p}$ Hamiltonian and used to calculate all intra- and inter-period transition rates due to interaction with phonons and electromagnetic radiation. Interaction with longitudinal acoustic phonons and electromagnetic radiation is included within the Fermi's golden rule, while electron LO-phonon interaction is included by considering polaron formation due to the strength of the interaction. A system of rate equations allows one to calculate the carrier distribution and evaluate parameters such as dark current and responsivity.

1.6.4 QDIPs in the THz

Huang et al. [58] reported a mid infrared QDIP with a double spectral response with a first band in the 3–13 μm region and another in the 20–55 μm, corresponding to 15–5.4 THz. At 150 K, the response in the terahertz range has a peak responsivity of 0.05 A / W and specific detectivity D* of 2×10^7 Jones, limited by the very big dark current in the device. The THz response is attributed to transitions between the higher excited states in the quantum dot and the continuum. Such claims are supported by the observed temperature behavior of the THz responsivity peak, which exhibits a temperature-activated behavior with a peak at 120 K.

This observation suggests the possibility of using QDIPs also for THz detection, even though the origin of the response is still not completely clear. Another nanostructure of interest, related to quantum dots, are the so-called quantum rings [59]. These are structures where electrons are confined into doughnut-like domains that behave like a rolled quantum wire. They are obtained by in-situ annealing of quantum dots during growth. Bandstructure calculations show that THz response is expected from such systems, which is and indeed observed in [60], where authors report a very broad THz spectral response. The devices exhibit extremely low dark current density due to the presence of resonant tunnelling barriers. Three prominent response peaks are observed at 6.5, 10, and 12.5 THz up to T = 120 K.

The quantum ring design was then optimized by the same authors in [61] by having only one bound state in the structure producing a response peak at 1.82 THz. An exceptional peak responsivity of 25 A/W, with a specific detectivity, D*, of 1×10^{16} Jones was measured at 5.2 K. Although the response vanishes completely at 12 K, the low temperature values are comparable to those of bolometers.

References

1. H. Schneider, H.C. Liu, *Quantum Well Infrared Photodetectors, Physics and Device Applications* (Springer, Berlin, 2007)
2. G. Bastard, *Wave Mechanics Applied to Semiconductor Heterostructures (les ditions de physique* (Les Ulis, France, 1988)
3. P. Harrison, *Quantum Wells, Wires and Dots: Theoretical and Computational Physics* (Wiley, New York, 1999)
4. C. Sirtori, F. Capasso, J. Faist, S. Scandolo, Nonparabolocity and a sum rule associated with bound-to-bound and bound-to-continuum intersubband transitions in quantum wells. Phys. Rev. B **50**, 8663 (1994)
5. T. Ando, A.B. Fowler, F. Stern, Electronic properties of two-dimensional systems. Rev. Mod. Phys. **54**, 437 (1982)
6. T. Unuma, M. Yoshita, T. Noda, H. Sakaki, H. Akiyama, Intersubband absorption linewidth in GaAs quantum wells due to scattering by interface roughness, phonons, alloy disorder, and impurities. J. Appl. Phys. **93**, 1586 (2003)
7. K.M.S.V. Bandara, D.D. Coon, O. Byungsung, Y.F. Lin, M.H. Francombe, Exchange interactions in quantum well subbands. Appl. Phys. Lett. **53**, 1931 (1988); erratum **55**, 206 (1989)

8. M. Helm, The basic physics of intersubband transitions, in *Intersubband Transition in Quantum Wells: Physics and Device Applications I. Semicon-ductors and Semimetals*, vol. 62, Chap. 1, ed. by H.C. Liu, F. Capasso (Academic, San Diego, 2000)

9. F. Castellano, R.C. Iotti, F. Rossi, Sequential multiphoton strategy for semiconductor-based terahertz detectors. J. Appl. Phys. **104**, 123104 (2008)

10. F. Castellano, F. Rossi, J. Faist, E. Lhuillier, V. Berger, Modeling of dark current in midinfrared quantum well infrared photodetectors. Phys. Rev. B **79**, 205304 (2009)

11. E. Lhuillier, I. Ribet-Mohamed, A. Nedelcu, V. Berger, E. Rosencher, Quantum transport in weakly coupled superlattices at low temperature. Phys. Rev. B **81**, 155305 (2010)

12. E. Lhuillier, E. Rosencher, I. Ribet-Mohamed, A. Nedelcu, L. Doyennette, V. Berger, Quantum scattering engineering of quantum well infrared photodetectors in the tunneling regime. J. Appl. Phys. **108**, 113707 (2010)

13. J.Y. Andersson, L. Lundqvist, Grating coupled quantum well infrared detectors, in *Long Wavelength Infrared Detectors*, Chap. 4, ed. by M. Razeghi (Gordon and Breach, Amsterdam, 1996)

14. S.S. Li, Metal grating coupled bound-to-miniband transition III-V quantum well infrared photodetectors, in *Long Wavelength Infrared Detectors*, Chap. 3, ed. by M. Razeghi (Gordon and Breach, Amsterdam, 1996)

15. S. Kalchmair, H. Detz, G.D. Cole, A.M. Andrews, P. Klang, M. Nobile, R. Gansch, C. Ostermaier, W. Schrenk, G. Strasser, Photonic crystal slab quantum well infrared photodetector Appl. Phys. Lett. **98**, 011105 (2011)

16. S.D. Gunapala, S.V. Bandara, J.K. Liu, J.M. Mumolo, D.Z. Ting, C.J. Hill, J. Nguyen, First demonstration of megapixel dual-band QWIP focal plane array, in *Proceedings of Sensors, 2009 IEEE*, 2009, pp. 1609–1612

17. H.C. Liu, C.Y. Song, A.J. SpringThorpe, J.C. Cao, Terahertz quantum-well photodetector. Appl. Phys. Lett. **84**, 4068 (2004)

18. H. Luo, H.C. Liu, C.Y. Song, Z.R. Wasilewski, Background-limited terahertz quantum-well photodetector. Appl. Phys. Lett. **86**, 231103 (2005)

19. F. Castellano, R.C. Iotti, F. Rossi, Improving the operation temperature of semiconductor-based terahertz photodetectors: a multiphoton design. Appl. Phys. Lett. **92**, 091108 (2008)

20. H. Schneider, H.C. Liu, S. Winnerl, C.Y. Song, M. Walther, M. Helm, Terahertz two-photon quantum well infrared photodetector. Opt. Express **17**, 12279–12284 (2009)

21. C.H. Yu, B. Zhang, W. Lu, S.C. Shen, H.C. Liu, Y.-Y. Fang, J.N. Dai, C.Q. Chen, Strong enhancement of terahertz response in GaAs/AlGaAs quantum well photodetector by magnetic field. Appl. Phys. Lett. **97**, 022102 (2010)

22. X.G. Guo, R. Zhang, H.C. Liu, A.J. SpringThorpe, J.C. Cao, Photocurrent spectra of heavily doped terahertz quantum well photodetectors. Appl. Phys. Lett. **97**, 021114 (2010)

23. X.G. Guo, Z.Y. Tan, J.C. Cao, H.C. Liu, Many-body effects on terahertz quantum well detectors. Appl. Phys. Lett. **94**, 201101 (2009)

24. M. Patrashin, I. Hosako, K. Akahane, Type-II InAs/GaInSb superlattices for terahertz range photodetectors. Proc. SPIE **8188**, 81880G (2011)

25. H. Machhadani, Y. Kotsar, S. Sakr, M. Tchernycheva, R. Colombelli, J. Mangeney, E. Bellet-Amalric, E. Sarigiannidou, E. Monroy, F.H. Julien, Terahertz intersubband absorption in GaN/AlGaN step quantum wells. Appl. Phys. Lett. **97**, 191101 (2010)

26. L. Gendron, M. Carras, A. Huynh, V. Ortiz, C. Koeniguer, V. Berger, Quantum cascade photodetector. Appl. Phys. Lett. **85**, 2824 (2004)

27. F.R. Giorgetta, E. Baumann, D. Hofstetter, C. Manz, Q. Yang, K. Köhler, M. Graf, InGaAs/AlAsSb quantum cascade detectors operating in the near infrared. Appl. Phys. Lett. **91**, 111115 (2007)

28. M. Tchernycheva, L. Nevou, L. Doyennette, F.H. Julien, E. Warde, F. Guillot, E. Monroy, E. Bellet-Amalric, T. Remmele, M. Albrecht, Systematic experimental and theoretical investigation of intersubband absorption in GaN?AlN quantum wells. Phys. Rev. B **73**, 125347 (2006)

29. A. Vardi, G. Bahir, F. Guillot, C. Bougerol, E. Monroy, S.E. Schacham, M. Tchernycheva, F.H. Julien, Near infrared quantum cascade detector in GaN/AlGaN/AlN heterostructures. Appl. Phys. Lett. **92**, 011112 (2008)

30. F.R. Giorgetta, E. Baumann, R. Théron, M.L. Pellaton, D. Hofstetter, M. Fischer, J. Faist, Short wavelength (4 μm) quantum cascade detector based on strain compensated InGaAs/InAlAs. Appl. Phys. Lett. **92**, 121101 (2008)

31. D. Hofstetter, F.R. Giorgetta, E. Baumann, Q. Yang, C. Manz, K. Köhler, Midinfrared quantum cascade detector with a spectrally broad response. Appl. Phys. Lett. **93**, 221106 (2008)

32. L. Gendron, C. Koeniguer, V. Berger, High resistance narrow band quantum cascade photodetectors. Appl. Phys. Lett. **86**, 121116 (2005)

33. M. Graf, N. Hoyler, M. Giovannini, J. Faist, D. Hofstetter, InP-based quantum cascade detectors in the mid-infrared. Appl. Phys. Lett. **88**, 241118 (2006)

34. F.R. Giorgetta, E. Baumann, M. Graf, L. Ajili, N. Hoyler, M. Giovannini, J. Faist, D. Hofstetter, P. Krötz, G. Sonnabend, 16.5 μm quantum cascade detector using miniband transport. Appl. Phys. Lett. **90**, 231111 (2007)

35. A. Buffaz, M. Carras, L. Doyennette, A. Nedelcu, X. Marcadet, V. Berger, Quantum cascade detectors for very long wave infrared detection. Appl. Phys. Lett. **96**, 172101 (2010)

36. A. Nedelcu, V. Guriaux, A. Bazin, L. Dua, A. Berurier, E. Costard, P. Bois, X. Marcadet, Enhanced quantum well infrared photodetector focal plane arrays for space applications. Infrared Phys. Technol. **52**, 412 (2009)

37. M. Graf, G. Scalari, D. Hofstetter, J. Faist, H. Beere, E. Linfield, D. Ritchie, G. Davies, Terahertz range quantum well infrared photodetector. Appl. Phys. Lett. **84**, 475–477 (2004)

38. E. Lhuillier, I. Ribet-Mohamed, E. Rosencher, G. Patriarche, A. Buffaz, V. Berger, M. Carras, Interface roughness transport in terahertz quantum cascade detectors. Appl. Phys. Lett. **96**, 061111 (2010)

39. P. Michler, *Single Quantum dots:? Fundamentals, Applications and New Concepts* (Springer, Berlin, 2003)

40. O. Stier, M. Grundmann, D. Bimberg, Electronic and optical properties of strained quantum dots modeled by 8-band k·p theory. Phys. Rev. B **59**, 5688 (1999)

41. D. Leonard, M. Krishnamurthy, C.M. Reaves, S.P. Denbaars, P.M. Petroff, Appl. Phys. Lett. **63**, 3203 (1993)

42. P. Atkinson, M.B. Ward, S.P. Bremner, D. Anderson, T. Farrow, G.A.C. Jones, A.J. Shields, D.A. Ritchie, Site-control of InAs quantum dots using ex-situ electron-beam lithographic patterning of GaAs substrates. Jpn. J. Appl. Phys. **45**, 2519 (2006)

43. A. Mohan, P. Gallo, M. Felici, B. Dwir, A. Rudra, J. Faist, E. Kapon, Record-Low inhomogeneous broadening of site-controlled quantum dots for nanophotonics. Small **6**, 1268 (2010)

44. J. Shao, T.E. Vandervelde, A. Barve, W.-Y. Jang, A. Stintz, S. Krishna, Enhanced normal incidence photocurrent in quantum dot infrared photodetectors. J. Vac. Sci. Technol. B **29**, 03C123 (2011)

45. K.W. Berryman, S.A. Lyon, M. Segev, Mid-infrared photoconductivity in InAs quantum dots. Appl. Phys. Lett. **70**, 1861 (1997)

46. S. Maimon, E. Finkman, G. Bahir, S.E. Schacham, J.M. Garcia, P.M. Petroff, Intersublevel transitions in InAs/GaAs quantum dots infrared photodetectors. Appl. Phys. Lett. **73**, 2003 (1998)

47. J. Phillips, K. Kamath, P. Bhattacharya, Characteristics of InGaAs quantum dot infrared photodetectors. Appl. Phys. Lett. **72**, 2020 (1998)

48. J. Urayama, T.B. Norris, J. Singh, P. Bhattacharya, Observation of phonon Bottleneck in quantum dot electronic relaxation. Phys. Rev. Lett. **86**, 4930 (2001)

49. A.V. Barve, S.J. Lee, S.K. Noh, S. Krishna, Review of current progress in quantum dot infrared photodetectors. Laser Photon Rev. **4**, 738 (2010)

50. R.V. Shenoi, R.S. Attaluri, A. Siroya, J. Shao, Y.D. Sarma, A. Stintz, T.E. Vandervelde, S. Krishna, Low-strain InAs/InGaAs/GaAs quantum dots-in-a-well infrared photodetector. AVS J. **26**, 1136 (2008)

51. L. Nevou, V. Liverini, F. Castellano, A. Bismuto, J. Faist, Asymmetric heterostructure for photovoltaic InAs quantum dot infrared photodetectors. Appl. Phys. Lett. **97**, 023505 (2010)

52. S. Krishna, Quantum dots-in-a-well infrared photodetectors. J. Phys. D: Appl. Phys. **38**, 2142 (2005)

53. E.-T. Kim, Z. Chen, A. Madhukar, Tailoring mid- and long-wavelength dual response of InAs quantum-dot infrared photodetectors using $In_xGa_{1-x}As$ capping layers. Appl. Phys. Lett. **79**, 3341 (2001)

54. G. Jolley, L. Fu, H.H. Tan, C. Jagadish, Effects of well thickness on the spectral properties of $In_{0.5}Ga_{0.5}As/GaAs/Al_{0.2}Ga_{0.8}As$ quantum dots-in-a-well infrared photodetectors. Appl. Phys. Lett. **91**, 193507 (2008)

55. S. Krishna, S. Gunapala, S. Bandara, C. Hill, D. Ting, Quantum Dot Based Infrared Focal Plane Arrays. Proc. IEEE **95**, 1838 (2007)

56. H. Lim, B. Movaghar, S. Tsao, M. Taguchi, W. Zhang, A.A. Quivy, M. Razeghi, Gain and recombination dynamics of quantum-dot infrared photodetectors. Phys. Rev. B **74**, 205321 (2006)

57. N. Vukmirovic, Z. Ikonic, I. Savic, D. Indjin, P. Harrison, A microscopic model of electron transport in quantum dot infrared photodetectors. J. Appl. Phys. **100**, 074502 (2006)

58. G. Huang, J. Yang, P. Bhattacharya, G. Ariyawansa, A.G.U. Perera, A multicolor quantum dot intersublevel detector with photoresponse in the terahertz range. Appl. Phys. Lett. **92**, 011117 (2008)

59. A. Lorke, R.J. Luyken, J.M. Garcia, P.M. Petroff, Growth and electronic properties of self-organized quantum rings. Jpn. J. Appl. Phys. **40**, 1857 (2001)

60. G. Huang, W. Guo, P. Bhattacharya, G. Ariyawansa, A.G.U. Perera, A quantum ring terahertz detector with resonant tunnel barriers. Appl. Phys. Lett. **94**, 101115 (2009)

61. S. Bhowmick, G. Huang, W. Guo, C.S. Lee, P. Bhattacharya, G. Ariyawansa, A.G.U. Perera, High-performance quantum ring detector for the 1–3 terahertz range. Appl. Phys. Lett. **96**, 231103 (2010)

Chapter 2
THz Bolometer Detectors

François Simoens

Abstract This chapter gives a review of bolometer use for THz detection, drawing specific attention to such sensors for THz imaging. First, the detection principles are presented highlighting the technological solutions that ensure the main functions of such thermal sensors. The two last sections are devoted to the review of technological approaches that are encountered depending on the operation temperature. Examples of developed cooled bolometer arrays are reviewed with emphasis on illustrating different approaches applied to perform the main bolometer functions. Finally, a focus is made on uncooled array of microbolometers, that is the field of R&D activity of the author.

Keywords Bolometer · Infrared detector · Terahertz detector · Thermal sensor · Antenna · THz imaging · Cooled bolometer · Uncooled bolometer

2.1 Bolometric Detection

This first section aims at giving for non-experts the basis of bolometer detectors that apply to detection in any spectral range, THz region included. The mechanism principles and figures of merit are introduced on the basis of a simplified model to put forward the main considerations and trade-offs that guide the design and choice of such sensors. For more details on the physics and technology of bolometers, the reader is given publications references [1, 2].

F. Simoens (✉)
CEA-LETI Minatec, 17 rue des Martyrs, 38054 Grenoble, France
e-mail: francois.simoens@cea.fr

M. Perenzoni and D. J. Paul (eds.), *Physics and Applications of Terahertz Radiation*,
Springer Series in Optical Sciences 173, DOI: 10.1007/978-94-007-3837-9_2,
© Springer Science+Business Media Dordrecht 2014

Fig. 2.1 Thermal detector principle

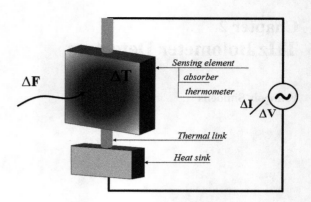

2.1.1 Bolometer Principle

The noun 'bolometer' is a composite word of greek origin, namely of 'bole' (beam, ray) and 'metron' (meter, measure). As a general principle, bolometers are thermal sensors that absorb electromagnetic radiation in an active area resulting in an increase of their temperature

As a general rule, thermal detectors gather a sensing element and a thermal link (Fig. 2.1). The sensing element itself combines two important parts that are an absorber—to collect the incident radiation or particles striking the sensor—and a temperature transducer (thermometer).

The thermal link isolates the sensing element from the heat sink (or reservoir temperature) to reduce heat losses but also provides mechanical support and electrical contacts between the bolometer and the read-out circuit.

The thermometer converts the resulting temperature variation into parameters that are measured electrically by the read-out circuit. So as illustrated by Fig. 2.2, thermal sensors detection makes use of a cascade of functions: electromagnetic absorber, thermometer, thermal link and read-out circuit. Some functions can be ensured by a common device. When a unique element acts as thermometer and absorber, thermal

Fig. 2.2 Thermal detector main functions

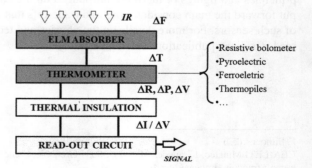

sensors are said to be monolithic; inversely, detectors made of a thermometer attached to a separate absorber are referred to as composite.

Different temperature transducers can be employed:

- Thermopiles are thermal sensors that exploit the Seebeck effect generated when two different electrically conducting materials are joined.
- Pyroelectric and ferroelectric sensors make use of electrical polarisation dependence in temperature.
- Thermistor bolometers in which the temperature variation is measured by a resistance change.[1] The change of resistance can be measured by passing a constant current I through the thermometer and then by observing the voltage drop across it, or vice versa.

Bolometers are incoherent sensors: unlike coherent detectors they provide only signal amplitude detection and therefore are not phase sensitive. In return, since thermal conversion is generally wavelength independent, the measured signals do not depend upon the spectral content of the radiant power, and hence bolometers can be very broadband detectors.

Another distinction can be mentioned: for thermal detectors, the excitations generated by the photons relax to a thermal distribution at a higher temperature (in the thermometer) before they are detected. In photon detectors, the non-thermal distribution of excited electrons is detected before it relaxes (e.g., to the conduction band).

2.1.2 Electro-Thermal Model

Electrical and thermal behaviours of bolometers have been extensively studied and discussed since the beginning of development of these detectors [1–3].

The main physical parameters that characterise a bolometer are (Fig. 2.3):

- $\alpha_{\text{opt}}(\lambda)$ and $C_{\text{th}}[\text{J/K}]$, respectively the spectral optical absorption efficiency (or absorptance of the detector) and the thermal capacity of the absorber;
- $R(T_b)[\Omega]$, the electrical resistance of the thermometer at the temperature T_b
- $G_{\text{th}}[\text{W/K}]$ and $R_{\text{th}} = 1/G_{\text{th}}[\text{K/W}]$, the heat conductance and resistance of the thermal link between the thermometer and the heat-sink at temperature T_{hs}.

---

[1] At this stage one has to distinguish between bolometer and calorimeter. The purpose of a bolometer is the measurement of a flux of particles or radiation, e.g. black-body radiation or infrared light. A flux or current is caused by single carriers of information, e.g. electrons in the case of electrical current.

By contrast, a calorimeter measures the energy of single particles absorbed in the detector. The purpose of a calorimeter is to measure the energy E of single quanta, e.g. X-ray photons, γ—quanta. This chapter will focus on such sensors applied to THz imaging and hence will only treat bolometer operation mode.

Fig. 2.3 Resistive bolometer
structure

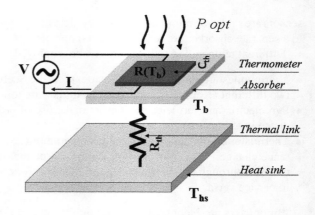

The thermal balance equation of the bolometer can be stated as

$$P_{\text{stocked}} = P_{\text{abs}} + P_{\text{bias}} - P_{\text{leak}} \tag{2.1}$$

where

- P_{abs} is the power issued from the optical absorption of external radiation or energy deposition. Considering a flux of particles or radiation P_{opt} imping-ing on the bolometer, the optical absorbed power can be expressed as $P_{\text{abs}} = \int \alpha_{\text{opt}}(\lambda) \, P_{\text{opt}}(\lambda) \mathrm{d}\lambda$. For clarity, this expression will be here simplified as, $P_{\text{abs}} = \eta \, P_{\text{opt}}$, η being the absorption efficiency with respect to an average incident optical flux.
- $P_{\text{bias}} = VI$ is the Joule heating induced by the read-out at constant current—or voltage—applied at the resistance, being respectively RI_{bias}^2 or V_{bias}^2/R.
- P_{stocked} is the power stocked in the bolometric thermal capacity: $P_{\text{stocked}} = C_{\text{th}}(\mathrm{d}T_b/\mathrm{d}t)$.
- P_{leak} is the power flowing from the bolometer to the constant temperature bath via the thermal link: this thermal leakage will be here expressed as $P_{\text{leak}} = -G_{\text{th}}(T_b - T_{\text{hs}})$. The effective thermal conductance might be much more complex to take into account radiative losses, dynamic effects and temperature dependence of the link materials especially when in the superconducting regime.

Conventional bolometers are operated in a vacuum package to minimise thermal loss between the bolometers and their surroundings through the surrounding gas convection. For example, high performance uncooled IR detector is integrated in specific packages that ensure typically 0.01 mbar vacuum [4]. So the power flow induced by this mechanism is usually negligible.

The heat flow budget can then be described by the following differential equation:

$$C_{\text{th}}\frac{\mathrm{d}(T_b - T_0)}{\mathrm{d}t} + G_{\text{th}}(T_b - T_{\text{hs}}) = \eta \, P_{\text{opt}} + VI \tag{2.2}$$

where T_o is a reference temperature that characterises the thermometer layer behaviour.

Static and dynamic behaviours can be described with an 'ideal' bolometer model, where the absorber and the thermometer are at the same temperature T_b, and stating that $T_o = T_{hs}$, and that the electrical resistance depends only on the temperature (non-ohmic effects are neglected).

Bolometer model without Joule heating

In a first approach, let us consider no Joule heating of the thermometer resistor and find a solution in the frequency domain with $P_{abs} = \eta\, P_{opt}(1 + e^{j\omega t})$ and $T_b - T_o = \delta T = \theta_{DC} + \tilde{\theta} e^{j\omega t}$, θ_{DC} and $\tilde{\theta}$ being the bolometer temperature variations with respect to T_o respectively in steady state and in small signals regimes.

Equation (2.2) can be expressed as:

$$j\omega C_{th}\left(\theta_{DC} + \tilde{\theta} e^{j\omega t}\right) = \eta\, P_{opt}\left(1 + e^{j\omega t}\right) - G_{th}\left(\theta_{DC} + \tilde{\theta} e^{j\omega t}\right) \qquad (2.3)$$

The steady-state regime is characterised by a temperature variation

$$\theta_{DC} = T_b - T_0 = \eta\, P_{opt}/G_{th} \qquad (2.4)$$

In the small signals regime, the solution of the frequency-dependent part of the temperature change equation is given by:

$$\tilde{\theta} = \frac{\eta\, P_{opt}}{G_{th}\left(1 + j\omega C_{th}/G_{th}\right)} \qquad (2.5)$$

When the radiation input power is absorbed by the detector, the temperature T_b initially increases with time at the rate $dT_b/dt = \eta\, P_{opt}/C_{th}$ and approaches $\lim_{t \to \infty} T_b = \theta_{DC} + T_0 = R_{th}\,\eta\, P_{opt} + T_0$ with the thermal time constant $\tau_{th} = R_{th}C_{th}$ (see Fig. 2.4).

This thermal behaviour is similar to a low frequency electrical circuit where the sensing element is equivalent to a capacitor and the thermal link is represented by a resistor. The time domain expression when the bolometer is exposed to a step of amplitude P_{opt} is:

$$T_b = T_0 + \eta\, P_{opt} R_{th}\left(1 - e^{-t/\tau_{th}}\right) \qquad (2.6)$$

When the radiation is turned off, the bolometer temperature relaxes back to T_0, exhibiting an exponential decay with characteristic decay time constant τ_{th}.

$$T_b = T_0 + \eta\, P_{opt} R_{th} e^{-t/\tau_{th}} \qquad (2.7)$$

The change of resistance dR caused by a temperature variation dT is characterised by the temperature coefficient of resistance TCR or defined at the reference

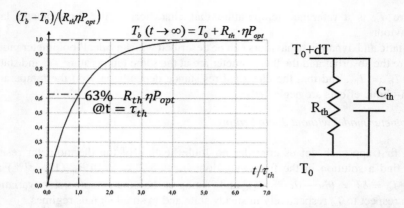

Fig. 2.4 The bolometer temperature evolution as a function of the normalised time with respect to when radiation is absorbed. At $t = \tau_{\text{th}}$, 63 % of the maximum temperature variation is reached, similarly to an equivalent R_{th}–C_{th} electrical low-frequency filter

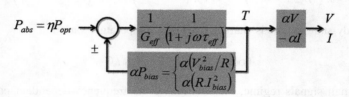

Fig. 2.5 An example of the bolometer temperature behaviour when submitted to an on-off radiation period applied every $10\tau_{\text{th}}$

temperature operating point T_0 by[2]

$$\alpha = \frac{1}{R}\frac{dR}{dT}\bigg]_{T_0} \tag{2.8}$$

expressed in [%/K].

The bolometer responds to a flux P_{opt} with a resistance change ΔR

$$\Delta R = \frac{\alpha R_0 R_{\text{th}}\,\eta\,P_{\text{opt}}}{1 + j\omega C_{\text{th}} R_{\text{th}}} \tag{2.9}$$

For high sensitivity, i.e. large change in resistance ΔR for a given impinging radiation P_{opt}, the parameters R_{th}, α, and η should be made as high as possible. On the other hand, in order to ensure a fast response, a small thermal constant $\tau_{\text{th}} = C_{\text{th}} R_{\text{th}}$ has to be achieved and consequently C_{th} and R_{th} should be minimised.

---

[2] TCR is sometimes defined as $\alpha = \frac{1}{R}\frac{dR}{dT}\big]_{T_0}$ with no units.

This simplistic model introduces a first key trade-off with respect to thermal detectors between sensitivity and response time. In particular, it shows the importance of the design of the thermal link.

Bolometer model with Joule heating

In a second step, let us consider P_{bias} the internal power caused by Joule heating of the thermometer resistor needed for the current or voltage measurement. With, $\delta T = T_b - T_0 = \theta_{\text{DC}} + \tilde{\theta} e^{j\omega t}$ the Eq. (2.3) can be rewritten as:

$$j\omega C_{\text{th}} \delta T + G_{\text{th}} \delta T = \eta \, P_{\text{opt}} \left(1 + e^{j\omega t}\right) + P_{\text{bias}} + \frac{\mathrm{d} P_{\text{bias}}}{\mathrm{d} T} \delta T \qquad (2.10)$$

Temperature change in the steady-state regime becomes

$$\theta_{\text{DC}} = T_b - T_0]_{t \to \infty} = \frac{\eta \, P_{\text{opt}} + P_{\text{bias}}}{G_{\text{th}} - \mathrm{d} P_{\text{bias}}/\mathrm{d} T} \qquad (2.11)$$

The small signals solution of the heat balance equation (2.10) is:

$$\tilde{\theta} = \frac{\eta \, P_{\text{opt}}}{(G_{\text{th}} - \mathrm{d} P_{\text{bias}}/\mathrm{d} T) + j\omega C_{\text{th}}} \qquad (2.12)$$

Depending on constant V_{bias} voltage or I_{bias} current bias conditions,

$$\mathrm{d} P_{\text{bias}}/\mathrm{d} T_{V_{\text{bias}}} = \frac{\mathrm{d} V_{\text{bias}}^2/R}{\mathrm{d} R} \frac{\mathrm{d} R}{\mathrm{d} T} = -\alpha P_{\text{bias}} \qquad (2.13)$$

or

$$\mathrm{d} P_{\text{bias}}/\mathrm{d} T_{I_{\text{bias}}} = \frac{\mathrm{d} \left(R I_{\text{bias}}^2\right)}{\mathrm{d} R} \frac{\mathrm{d} R}{\mathrm{d} T} = +\alpha P_{\text{bias}} \qquad (2.14)$$

Then

$$\tilde{\theta} = \frac{\eta \, P_{\text{opt}}}{G_{\text{eff}} \left(1 + j\omega \tau_{\text{eff}}\right)} \qquad (2.15)$$

Where and $G_{\text{eff}} = G_{\text{th}} \pm \alpha P_{\text{bias}}$ and $\tau_{\text{eff}} = C_{\text{th}}/G_{\text{eff}}$. This expression transduces an electro-thermal feedback (ETF) that is analogous to the electrical feedback with a loop gain $L = \pm \alpha P_{\text{bias}}/G_{\text{th}}$ (refer to Fig. 2.6).

The current or voltage changes as a function of the temperature are such that:

$$R_{I/T} = \frac{\partial I}{\partial T} = \frac{\partial \left(V_{\text{bias}}/R\right)}{\partial T} \frac{\partial R}{\partial T} = -\alpha \mathrm{I} \qquad (2.16)$$

$$R_{V/T} = \frac{\partial V}{\partial T} = \frac{\partial \left(R I_{\text{bias}}\right)}{\partial T} \frac{\partial R}{\partial T} = \alpha \mathrm{V} \qquad (2.17)$$

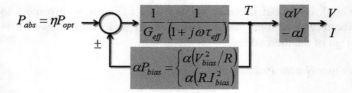

Fig. 2.6 A schematic diagram of the electro-thermal feedback principle

The change in the current or voltage drop across the thermometer as a consequence of the absorption of radiant power P_{opt} defines the bolometer responsivity.

For a constant V_{bias} voltage bias condition:

$$\mathbf{R}_{I/P_{abs}} = \frac{\partial I}{\partial P_{abs}} = \mathbf{R}_{I/T}\frac{\partial T}{\partial P_{opt}} = \frac{-\alpha I\,\eta}{G_{th}}\frac{1}{(1+L)}\frac{1}{(1+j\omega\tau_{eff})} \qquad (2.18)$$

For a constant I_{bias} current bias condition:

$$\mathbf{R}_{V/P_{abs}} = \frac{\partial V}{\partial P_{abs}} = \mathbf{R}_{V/T}\frac{\partial T}{\partial P_{abs}} = \frac{\alpha V\,\eta}{G_{th}}\frac{1}{(1+L)}\frac{1}{(1+j\omega\tau_{eff})} \qquad (2.19)$$

where $\tau_{eff} = \tau_{th}/(1+L)$ $L = \pm\alpha P_{bias}/G_{th}$.

The electro-thermal feedback (ETF) modifies the effective thermal conductance and time constant of the link to the heat sink according to the TCR sign and measurement conditions:

	Constant V_{bias} $G_{eff} = G_{th} + \alpha P_{bias}$	Constant I_{bias} $G_{eff} = G_{th} - \alpha P_{bias}$
$\alpha > 0$ Case of superconductor (or metallic) bolometer	$G_{eff} > G_{th} \rightarrow \mathbf{R}_{1/P_{abs}} \downarrow$ $\tau_{eff} < \tau_{th}$	$G_{eff} < G_{th} \rightarrow \mathbf{R}_{V/P_{abs}} \uparrow$ $\tau_{eff} > \tau_{th}$
$\alpha < 0$ Case of semiconductor bolometer	$G_{eff} < G_{th} \rightarrow \mathbf{R}_{1/P_{abs}} \uparrow$ $\tau_{eff} > \tau_{th}$	$G_{eff} > G_{th} \rightarrow \mathbf{R}_{V/P_{abs}} \downarrow$ $\tau_{eff} < \tau_{th}$

Therefore, ETF also intervenes in the trade-off between the sensitivity and the response time when designing a bolometer. The sensitivity improvement would argue for biasing such that $G_{eff} < G_{th}$. However, in addition to the degradation of the response time, this option exhibits instability of the feedback loop that may end up in thermal runaway and consequently in the degradation of the bolometer. Actually, $G_{eff} = 0$ constitutes a pole of this loop and it has to remain positive for stable operation. For example, this condition is always true for a positive (the case of a superconductor bolometer) and a constant voltage bias. But if the voltage changes are measured in such temperature transducers, the current has to be limited such that G_{eff} always stays positive.

Semiconductor bolometers are most often operated in a constant-current mode, while for a superconducting thermometer constant-voltage operation is favoured. However, in addition to these considerations, the read-out circuit design criteria—impedance matching, multiplexing technology,. . .—may greatly intervene in the choice of the bias conditions. For example, most of commercialised uncooled semiconductor bolometers are read in current with a pulsed voltage bias, though negative TCR exposes to a potential thermal runaway.

2.1.3 Sources of Noise

Various sources of noise have detrimental effects on the bolometer figures of merit. A very detailed analysis is given in [3].

In order to simplify the expressions here, a constant voltage bias and measurement of the current change across the bolometer resistance are assumed.

Phonon noise

Thermal fluctuation noise—or phonon noise—arises from the random exchange of energy between the bolometer and the heat sink. Assuming an equilibrium distribution of phonons,[3] classical thermodynamic statistical mechanics analysis gives the mean square magnitude of the energy fluctuations as $\langle \Delta E^2 \rangle = k_B T_b^2 C_{\text{th}}$, where T_b is the operating temperature of the bolometer.

The mean square temperature fluctuation is:

$$\langle \Delta T \rangle^2 = \frac{\langle \Delta E \rangle^2}{C_{\text{th}}^2} = \frac{k_B T^2}{C_{\text{th}}^2} = \frac{1}{2\pi} \int_0^\infty S_T(\omega)\, d\omega \qquad (2.20)$$

Since the temperature and the power spectral densities, respectively $S_{T\_\text{phonon}}(\omega)$ and $S_{T\_\text{phonon}}(\omega)$ due to phonon noise are linked as follows[4]:

$$S_{T\_\text{phonon}}(\omega) = S_{P\_\text{phonon}}(\omega) \left(\frac{dT}{dP_{\text{abs}}} \right)^2 \qquad (2.21)$$

So $S_{T\_\text{phonon}} = 4k_B T^2 / G_{\text{eff}} [\text{A}^2/\text{Hz}]$ and $S_{T\_\text{phonon}} = 4k_B T^2 / G_{\text{eff}} [\text{W}^2/\text{Hz}]$.

$$S_{(I_\text{phonon})}(\omega) = (R_I / P_{\text{abs}})^2 4 k_B T^2 G_{\text{eff}} \qquad (2.22)$$

---

[3] Energy carriers need not be restricted to phonons, electrons like in hot electron bolometers, photons or quasiparticles being possibly exchanged. But in this chapter, only phonons will be considered.

[4] The power density can be computed by $\frac{k_B T^2}{C_{\text{th}}} = \frac{1}{2\pi} \int_0^\infty S_P(\omega) \left(\frac{1}{G_{\text{eff}}\sqrt{1+\omega^2 \tau_{\text{eff}}^2}} \right) d\omega = \frac{1}{2\pi} \frac{\pi S_P(\omega)}{G_{\text{eff}}^2 2\tau_{\text{eff}}}$.

This noise source is often the dominant intrinsic noise source that limits the sensitivity of a bolometer in practice.

Johnson noise

The cause of Johnson noise is the random fluctuation of electrons in the resistance of the bolometer. This dissipation fluctuation is temperature dependent, and consequently the noise is reduced at lower temperatures. The original Johnson noise current spectral density is then

$$S_{I\_J} = 4k_B T_b / R \qquad [\text{A}^2/\text{Hz}] \tag{2.23}$$

If constant current bias is applied, the voltage noise density is:

$$S_{V\_J} = 4k_B T R \qquad [\text{V}^2/\text{Hz}] \tag{2.24}$$

1/f noise

This noise depends strongly on the properties of the thermometer and on the coupling between the bolometer layer interfaces. Excess low-frequency noise caused by various sources may also be identified as $1/f$ noise.

In particular, this noise can be seen as low frequency fluctuations of the electrical conductivity that follow the Hooge law with k, a unit-less factor. The corresponding current noise spectral density is then

$$S_{I\_1/f} = \frac{kV^2}{fR^2} \qquad \left[A^2/\text{Hz}\right] \tag{2.25}$$

This noise pushes the research of optimum detector materials and good electrical contacts. This $1/f$ noise is of particular significance for room temperature bolometers.

Photon noise

Photons radiated from the detector and those impinging on it also carry energy. Radiative exchange of energy takes place with the environment of the detector, e.g. the walls of the enclosure. This noise source can be neglected if the surroundings of the detector are sufficiently cooled.

A noise source which cannot be avoided arises from background radiation in front of which a radiative source is to be observed. A detector is called background limited (BLIP) when this type of photon noise dominates.[5] Most fundamental noise comes from the source itself.

---

[5] For example, this is the case of Ge-bolometers operated at a temperature of 300 mK.

Load noise

A load resistor R_L for the thermometer also contributes to Johnson noise, as does the thermometer. This noise is usually made negligible by using a value for R_L much higher than the bolometer resistor ($R_L \gg R_b$).[6] It is further neglected.

Excess noise

This term arises from various noise sources associated with the environment. There could be temperature fluctuations in the temperature reservoir, vibration, external electrical interference, electromagnetic environment or microphony of the detector. It is important to reduce these noise sources to a level well below the unavoidable noise sources.

Further, to implement a specific kind of detector bias and read-out scheme, electrical components are necessary which cause an increase of the system noise. Of course, the transfer functions of the various components present in the circuitry of the detector and preamplifier have to enter into the noise calculation. It is thus always a trade-off between signal-to-noise ratio and other requirements imposed on the detector read-out.

2.1.4 Figures of Merit

Since bolometers are used in such a wide variety of applications, an attempt to compare detectors may be sometimes confusing. Five main figures of merit are commonly employed to characterise the performance of bolometers, i.e. the spectral response $\alpha_{opt}(\lambda)$, the response time defined by the thermal constant τ_{th}, the responsivity $\boldsymbol{R}$, the noise equivalent power and the noise equivalent temperature difference *NETD*.

Spectral response: $\alpha_{opt}(\lambda)$ transduces the dependence of the pixel output signal upon the wavelength of the incident radiation for a constant incident intensity per wavelength interval falling onto the pixels. The units are dimensionless. Depending on the targeted application, the bolometers spectral absorption has to be tuned on the corresponding range. The following Figs. 2.7 and 2.8 show examples of spectral response of infrared and terahertz bolometers, respectively.

Response time: referring to Sect. 2.1.2, the bolometer response time is characterised by its effective thermal time constant:

$$\tau_{eff} = C_{th}/G_{eff} \tag{2.26}$$

Responsivity: it is the ratio between the electrical output signal and the impinging power (refer to 2.1.2).

---

[6] And by cooling to very low temperature for cooled operation bolometers.

Fig. 2.7 An example of the
spectral response of a typical
uncooled microbolometer
optimised to sense thermal
infrared radiation in the 8–
14 μm atmospheric window

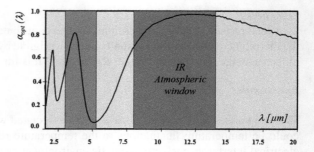

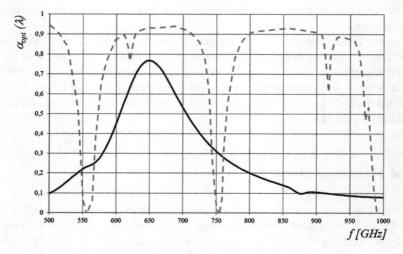

Fig. 2.8 An example of the simulated spectral response (*plain line*) of an antenna–coupled
microbolometer designed at CEA-Leti to sense terahertz radiation in the 650 GHz atmospheric
window (*dashed line*)

Noise equivalent power (*NEP*): the NEP of a bolometer is defined as the power that
generates a signal output that is equal to the root-mean-square (rms) noise output
in a 1 Hz bandwidth. In other words, the NEP is the signal power which causes a
signal-to-noise ratio (SNR) equal to one.

$$\text{NEP} = \frac{v_n}{R_v} = \frac{i_n}{R_i} \tag{2.27}$$

The unit of NEP is the Watt. The NEP is also quoted for a fixed reference bandwidth, which is often assumed to be 1 Hz. This "NEP per unit bandwidth" has units
of Watts per square root Hertz $[\text{W}/\text{Hz}^{1/2}]$. Often, its inverse, the detectivity D is
quoted where $D = \text{NEP}^{-1}$.

Other authors consider the specific detectivity D^* (pronounced "dee-star") defined
as the root-mean-square signal-to-noise ratio of a 1 Hz bandwidth per unit rms incident radiant power per square root of bolometer active area A:

$$D^* = D\sqrt{A} \qquad \left[\mathrm{cm\,Hz}^{1/2}/\mathrm{W}\right] \qquad (2.28)$$

D* provides the possibility to compare bolometer pixels of the same type but with different pixel areas[7]. This parameter assumes that the noise is white in the frequency range of interest. This may not hold for bolometers with thermometers which exhibit high $1/f$ noise.

Noise equivalent temperature difference (NETD): The NETD is defined by the change in temperature of a large blackbody in a scene being viewed by a thermal imaging system which will cause a change in the signal equal to the rms noise level in the output of a pixel upon which part of the blackbody is imaged.

In the THz range, this figure of merit is useful for passive THz imaging. It is related to the electrical NEP of the detector through:

$$\mathrm{NETD} = \frac{\mathrm{NEP}}{\eta\, k_B \Delta\nu \sqrt{2\tau_{\mathrm{int}}}} \qquad (2.29)$$

where the validity of Rayleigh-Jeans law is assumed ($h\nu \gg k_B T$), η is the coupling efficiency of the optical radiation, $\Delta\nu$ is the RF bandwidth and τ_{int} is the post-detection integration time.

NETD is one of the most important performance parameters for infrared imaging systems, especially because it is quite directly characterised by exposing the bolometer to the radiation of two side-by-side blackbody emitters of large lateral extent set at two different temperatures. But such characterisation is not easy in the THz spectrum where black-body radiation is very poor compared to the thermal emission in the infrared range.

2.1.5 General Discussion

First, neglecting Joule heating and considering only the dominant noise sources that are typically phonon, Johnson and $1/f$ uncorrelated noises, the total current noise spectral density can be written as follows:

$$S_I = 4k_B T^2 G_{\mathrm{eff}} R_I^2 + 4k_B T_b/R + \frac{kV^2}{fR^2} \qquad \left[\mathrm{A}^2/\mathrm{Hz}\right] \qquad (2.30)$$

S_I is converted to a noise equivalent power simply by dividing this total noise density by the current responsivity R_1 of the detector:

---

[7] The figure D^* is useful in cases where the dominant noise source is the black-body radiation which exhibits an area dependence of $\sqrt{A}$. However, not all detectors, especially those operating at very low temperature, show such noise behaviour.

Fig. 2.9 An example of the computed NEP versus bias voltage

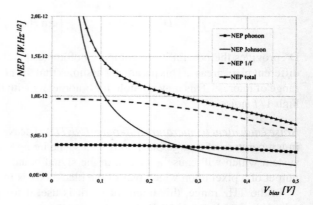

$$\text{NEP}^2 = \text{NEP}^2_{\text{phonon}} + \text{NEP}^2_{\text{Johnson}} + \text{NEP}^2_{1/f} \tag{2.31}$$

$$\text{NEP}^2 = 4k_B T_b^2 G_{\text{eff}} + \frac{4k_B T_b}{R R_I^2} + \frac{k V_{\text{bias}}^2}{f R^2 R_I^2} \tag{2.32}$$

In order to illustrate the NEP considerations the bolometer designer has to deal with, let us consider an uncooled amorphous silicon bolometer that exhibits the following typical values: $\alpha = -2.5\,\%/\text{K}$, $k = 6.5 \times 10^{-12}$, $G_{\text{th}} = 3.10^{-8}\,\text{W/K}$, $R = 500\,\text{k}\Omega$, and $\tau_{\text{th}} = 10\,\text{ms}$. Supposing a 100 nW incident optical power absorbed with ideal unity efficiency and chopped at a frequency of 10 Hz, it is possible to compute the NEP contributions for each type of noise.

Figure. 2.9 shows the computed NEP versus V_{bias} voltage bias. This specific case shows two main tendencies. First, $\text{NEP}_{\text{Johnson}}$ and $\text{NEP}_{1/f}$ decrease as V_{bias} increases. This result is in accordance to the expected responsivity enhancement caused by ETF when $\alpha < 0$ and a constant voltage bias is applied.

Moreover, this computation shows that the thermal insulation between the microbolometer and the heat sink determines the lower limit in terms of NEP (in this example 0.4 pW/HZ$^{1/2}$ for $T_b = 300\,\text{K}$). This value is always below the contribution of $1/f$ noise, which is the limiting factor in such examples.

Beyond 0.5 V, the positive thermal feedback will provoke the destruction of the bolometer as the previous Fig. 2.10 plots the temperature runaway caused by the bias voltage increase.

To sum up, the design of bolometers with a high sensitivity deals with a number of design features and trade-offs. Some of the most important parameters are:

- High absorption of the radiation including a large absorbing area,
- Low thermal conductance between the bolometer and its surrounding,
- Sufficiently low bolometer thermal time constant.
- Bolometer temperature sensing material with a high temperature coefficient of resistance (TCR) and low $1/f$ noise properties.

This last criterion is directly related to the choice of appropriate thermistor material. The following subsection presents an overview of the main materials that are used in THz bolometers.

2.1.6 Thermometer Materials

In this subsection, some of the different types of temperature transducers are discussed [5]. The most commonly used are the metal, the thermistor, the semiconductor and the superconducting bolometer. They differ significantly in their temperature dependence of resistance as shown in Fig. 2.11.

Metal strips

Metals strips were used for the first bolometer invented by the American scientist Langley [6] in 1880 for solar observations in the infrared range. It consisted in a Wheatstone bridge used along with platinum foils.

Metals exhibit in a first approximation, a linear dependence of the resistance on temperature. Thus, a change in temperature ΔT causes a variation in resistance. $\Delta R = \alpha\, R_0\, \Delta T$.

Fig. 2.10 The computed temperature variation versus the bias voltage

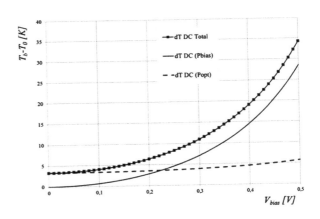

Fig. 2.11 The three main materials for bolometer

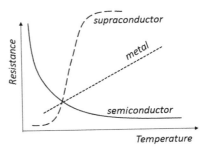

Fig. 2.12 The first bolometer as invented by S.P. Langley in 1880 [6]

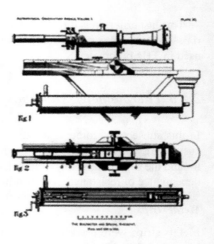

The specific heat capacity of normal metal strips is comparatively high. Hence, to achieve high detector sensitivity the bolometers need to be small. The good mechanical properties of nickel and platinum made them preferred materials for bolometers. Also of interest is bismuth because of its low electronic heat capacity and titanium due to its low thermal conductance.

In such metals, the temperature coefficient is proportional to T^{-1}. Thus, lower operating temperatures are more favourable if high sensitivity is a concern. The TCR is unfortunately very low at room temperature, typically ~ 0.3 /K.

Thermistors

Thermistors consist of metal oxides (like manganese, cobalt or nickel). A temperature increase of the film causes an increase of the density of free charge carriers, and consequently a resistance reduction. The temperature coefficient is about $\alpha \sim 5$ /K.

TCR exhibits a temperature dependence in T^{-2}. Sensitivities close to that value in semiconductor thermometer can be achieved.

Semiconductor thermometers

These bolometers benefit from a long history with the first developments of infrared detectors.

At low temperature, semiconductor thermometers can be very sensitive. For example, a germanium single crystal doped with gallium operated at a temperature of $T = 2$ K may perform a TCR $\sim -2 K^{-1}$. Since the days of early cooled semiconductor thermometers, improvements have been made and values of between -4 and $10 K^{-1}$ can easily be produced. Intrinsic semiconductor thermometers exhibit an exponential temperature dependence of the resistance:

$$R(T) = R_0 \exp(T/T_0)^B \approx -(B/T_0)(T_0/T)^{B+1}$$

where B is a coefficient characteristic of the material. Intrinsic or lightly doped semiconductor thermometers exhibit a coefficient $B \sim 1$. They are not useful at low temperatures because of their excessively high resistance which is difficult to read-out (excess of noise and impedance mismatching). For cooled operation, heavily doped semiconductors are used, and preferably with a doping suitable for metal–semiconductor transition behaviour of such materials. The electron transport is governed by a variable-range hopping mechanism leading to a parameter B close to 0.5.

Today for bolometers operating below 4 K, mainly silicon (Si) and germanium (Ge) are employed. To overcome technological difficulties in achieving uniform doping with classical processes, specific techniques have been developed, based on ion implantation in Si or irradiation of pure Ge with neutrons[8] followed by thermal annealing (realizing Neutron Transmuted Doped Germanium or NTD-Ge).

Heavily doped and compensated semiconductors conduct by a hopping process that yields the resistance $R(T) = R_0\exp(T/T_0)^p$, where R is the resistance at the temperature T, and T_0, R_0 are the constants which depend on the doping and on the thermistor dimensions. p is a constant that depends on the semiconductor material. The resulting TCR is negative $A = \frac{T}{R}\frac{dR}{dT}$. Typically, $|A| = 6\cdots15$ for temperature ranging from 20 to 300 mK.

Doped semiconductors exhibit a high specific heat which is detrimental for detectors of high sensitivity. With modern technology, however, it is possible to dope only a very restricted area of the absorber.

One interesting alternative to Si or Ge is the semiconducting YBaCuO that can have TCR values between 2.8 and 4 %/K.

At room temperature, the most common uncooled bolometer temperature sensing materials are semiconductors, i.e. vanadium oxide (VOx), amorphous silicon ($\alpha-$Si) and silicon diodes [7]. Such materials exhibit a typical TCR in the range -2 and -3 %/K.

As illustrated in the previous Sect. 2.1.5, the main limitation in sensitivity comes from the $1/f$ noise. The $1/f$ noise constant is a material parameter that can vary by several orders of magnitude for different materials. Mono-crystalline materials exhibit significantly lower $1/f$ noise contribution as compared to amorphous or poly-crystalline materials.

The vanadium oxide thin films are used today in a large variety of bolometer products. There are many phases in vanadium oxides, such as VO_2, V_2O_5 and V_2O_3.

Bolometers made of $\alpha-$Si can consist of very thin membranes, which allow for a low thermal mass. Consequently, $\alpha-$Si bolometers can benefit from a low thermal conductance while maintaining a fixed bolometer time constant. The relatively lower electronic performance of low-temperature $\alpha-$Si devices could be compensated by the cheaper production, for future, ultra-low-cost, high-volume applications.

In bolometer silicon diodes, the radiation absorption generates a voltage change of a pn-junction or Schottky barrier junction. Such transducers perform a TCR close to $s \sim 0.2$ %/K that is an order of magnitude lower than the other semiconductor

---

[8] NTD converts ^{70}Ge to ^{71}Ga (acceptor) and ^{74}Ge to ^{75}As (donor).

devices. In return, Si diodes can potentially be manufactured in standard CMOS lines.

Superconducting phase transition thermometers

Superconducting phase transition thermometers exploit the steepness of the transition between the superconducting and the normal-conducting phases. The electrical resistance drops abruptly when the material is cooled through its superconducting transition temperature named T_c. In this conductivity transition the resistance changes dramatically over the transition temperature range resulting in a TCR of several hundred per Kelvin. Furthermore, since the specific heat of most materials decreases significantly with temperature, the time constant τ_{th} of the bolometer shortens and thus faster detectors as compared to room-temperature detectors can be built. Bolometers integrating such thermometer are usually named as transition-edge sensors (TES) [8].

Changes in temperature transition can be set by using a bilayer film consisting of a normal material and a layer of superconductor. Such designs enable diffusion of the Cooper pairs from the superconductor into the normal metal and make it weakly superconducting—this process is called the proximity effect.

At low temperature, superconducting phase transition thermometers using low T_C superconductors such as Al, Ta, Sn, Nb or NbN are developed for THz sensing. The rapid progress in semiconductor technology, however, furnished semiconductor thermometers which exhibited comparable sensitivity. As read-out electronics for semiconductor thermistors are generally less complex than for superconducting phase transition thermometers, the choice of instruments was very often in favour of the semiconductor bolometer.

Superconducting bolometers are already applied in a wide field of research. Temperature transducers involving superconductivity can be built in a great variety, in particular low-T_c, and high-T_c bolometers.

Hot Electron Bolometers

The idea of hot electron bolometer (HEB) is to use electrons as a subsystem in semiconductors or superconductors that interacts with the lattice phonons. Since the electron thermal capacity is much lower than the lattice, the heat induced by radiation at the sensitive element level can quickly escape through the thermal link, and thus ensures a faster response [9, 10]. The processes of hot-electron effect below the transition temperature T_c that takes place in superconducting thin films result from electron–phonon interactions where energy relaxation times and times of escape into the substrate compete.

The first bolometer with hot electron bolometer used low temperature bulk n-InSb. Other semiconductor materials used nowadays to perform with $\tau_{th} \sim 10^{-7}$s, and consequently fast response. Thus, these detectors are suitable for incoherent systems.

Different superconducting materials (Nb, NbN, . . .) are sensitive in a wide range of spectra and are much faster compared to bulk semiconductor bolometers at, reaching a thermal time constant in the order of a picosecond. Such materials are very attractive to make HEB mixers for high sensitivity and low local oscillator (LO) power ($1 <$ μW) terahertz receivers [11].

High T_cBolometers

Thermometers can favourably be made with material compounds like yttrium, barium, copper and oxygen (YBCO) that exhibit high-T_c detection [12, 13]. Their strong interest is that they can be operated at liquid-N_2 temperatures achievable with more compact and simpler cryogenic systems—like Sterling modules—in comparison to He-cooled devices.

In return, their performance is orders of magnitude worse than those operating at liquid-He temperature, but better than those operating at room temperature. Moreover, since they are made of compounds and are grown on specific substrate materials (Si, sapphire, ZrO_2, $LaAlO_3$ or SiN...) fabrication of high- bolometers is more delicate than the fabrication of the classical low-T_c.

Most of the above-mentioned materials can either be used at the same time as the radiation absorber and for the thermometer functions (monolithic sensor). Alternatively, they can be dedicated to temperature transducing, whereas a separate element ensures absorption. Nowadays, composite bolometers are the predominately used type of detector: they offer the advantages of fabricating detectors with large active absorbing area or volume and freedom in separate optimisation of critical bolometer parameters, like thermal conductivity, heat capacity and absorptivity.

Composite design approach is especially relevant for terahertz bolometers, where the size of the optical absorber element has to follow the wavelength that is significantly larger than in the infrared range. The following section gives example of developed cooled bolometer arrays and how different technologies are employed to carry out separately the main bolometer functions described in Sect. 2.1.1 (electromagnetic absorber, thermometer, thermal link). Intentionally, only examples of multi-element bolometer sensors used for THz direct detection are presented. But, as mentioned previously, cooled bolometer utilising the hot electron phenomenon and transition edge sensors are suitable for heterodyne detection.

2.2 Cooled THz Bolometers

2.2.1 General Considerations

In reference to the NEP equation derived in the previous section, superconducting bolometers [14, 15] offer a clear advantage in comparison to uncooled bolometers because of their low *NEP* due to the low thermal noise and high TCR. Research on superconducting bolometers begun in 1938 at the Johns Hopkins University and was motivated by the idea that exploitation of superconductivity could lead to faster and more sensitive bolometers. The most sensitive THz detectors are cryogenic bolometers with noise levels down to the fW $/\sqrt{\text{HZ}}$ -level.

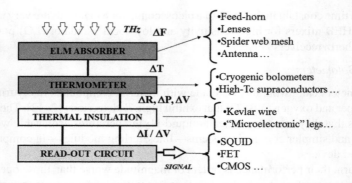

Fig. 2.13 A (not exhaustive) list of technologies employed in cooled THz bolometers to ensure the main thermal detector main functions

Fig. 2.14 A photograph of
the MAMBO2 Max-Planck
Institute bolometer array
http://www.mpifr-bonn.mpg.
de/div/bolometer

In return, the drawbacks of cryogenic operation are the difficulty in achieving appropriate thermal insulation, mechanical and electrical interfaces with the environment. For example, practical approaches to multiplexing many such bolometers to one amplifier (JFETs mostly) do not exist and current arrays are limited to a few hundred pixels with individual read-out channels. These difficulties have motivated technological developments combining cleverly the cooled THz bolometer functions (absorber-thermometer-thermal links-read-out circuits) in order to optimise its figures of merit.

The forthcoming subsection will give examples of these technologies developed for cooled bolometers.

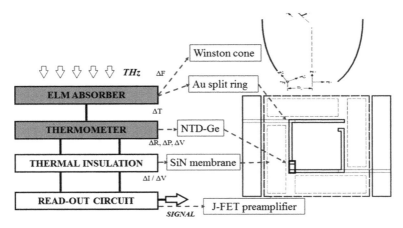

Fig. 2.15 The MAMBO2 bolometer with the decomposition of the functions

2.2.2 Example of Cryogenic Bolometers

The MAMBO and MAMBO-2 two are successive versions of millimeter bolometer focal plane arrays (FPA): they are representative examples of composite bolometer associated to feed-horns antenna for the radiation collection [16]. Developed by the 'Max-Planck Institut' for the IRAM 30 m telescope on Pico Veleta (Spain), these FPA contain respectively 37 and 117 pixels sensitive at 1.2 mm-wavelength and operated at a temperature of 300 mK.

As shown in Fig. 2.15, each pixel consists of a freestanding 3.4×3.4 mm^2 membrane of Silicon Nitride (SiN) that provides low thermal conductivity. On this membrane, a "split ring" of Gold achieves spatially uniform absorption of the radiation and a (NTD) Germanium thermometer is attached. These arrays make use of specific feed-horns named Winston cones to maximize the collection of incoming rays. The preamplifiers are based on low-noise junction field-effect-transistors (FETs) operating at 300 K.

The following example reviews the TES bolometer (Fig. 2.16) developed by the Berkeley group that illustrates the use of planar antennas associated to the bolometer on the detector chip to provide coupling to free space waves [17].

217 GHz double-slot dipole antennas are combined with a silicon hyper-hemispherical lens. One of the main actions of such lens coupled to the antenna is to synthesise an infinite dielectric, and hence it eliminates the substrate modes (it is often named "substrate lens") and narrows the beam for better optimal match to optics with virtually no spherical aberration or coma.

The power dissipated in the resistor is measured with a Transition-Edge Sensor (TES) made from an aluminum/titanium bilayer. Varying the thickness of the bilayer allows tuning of the superconducting critical temperature to optimise the detector performance. The load resistor and TES are in good thermal contact on a leg-isolated

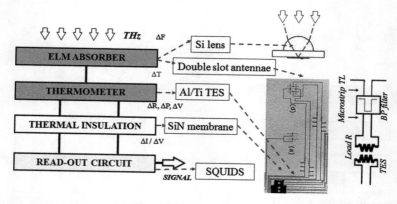

Fig. 2.16 The prototype of TES bolometer coupled to dipole planar antennas developed at Berkeley

silicon nitride substrate. This provides the thermal insulation needed for the required sensitivity.

The read-out of resistance change is ensured by a superconducting quantum inter-ference device (SQUID) that exploits the Josephson effect. This effect relates to a super-current that can flow between two pieces of superconductor separated by a thin layer of insulator. In SQUIDs, two Josephson junctions are connected in a loop and the SQUID output voltage is biased in such a way that very small changes can be detected (it acts as a low input impedance amplifier).

Another interesting approach consists in using micromesh as the absorber in combination to a feed-horn. Such devices—often named spiderweb meshes—minimise the suspended mass and heat capacity of the bolometers.

As an example [18, 19], feed-horn coupled bolometer arrays sensitive at 350 μm have been developed for the SPIRE instrument of the Herschel Space Observatory by JPL Caltech. Each array contains 139 pixels (size: $10 \times 100 \times 300 \mu m$) made of NTD Ge thermometers attached to a silicon nitride micromesh (= absorber) with In bump bonds. The bolometers are coupled to the telescope beam by conical feedhorns located directly in front of the detectors (see Fig. 2.17).

The achieved performance is a, NEP $= 1.5 \times 10^{-17} W/HZ^{1/2}$, $\tau_{th} = 11$ ms at 300 mK and a NEP $= 1.5 \times 10^{-18} W/HZ^{1/2}$, $\tau_{th} = 1.5$ ms, at 100 mK.

An example of a fully planar technology based on TES bolometers is given by the remarkable work initiated at NIST and pursued in collaboration with VTT (Fig. 2.18). Broadband antenna-coupled superconducting TES micro-bolometers have been assembled in linear arrays [20].

Each pixel consists of a Nb bridge encapsulated in SiO_2, suspended above a Si substrate at the feed-point of a lithographic circular spiral antenna. This antenna is designed to cover the 0.075–1.02 THz bandwidth. The suspension of the Nb film over an etched cavity ensures thermal insulation from the heat sink. Similarly to the previously example of the Berkeley bolometers, logarithmic antennas are coupled to substrate lenses to avoid substrate mode losses and to provide better optical matching.

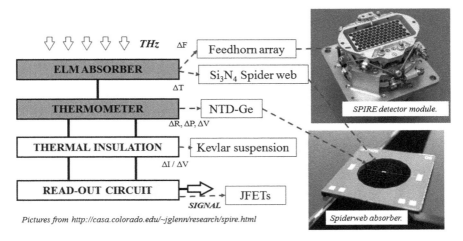

Fig. 2.17   JPL-Caltech bolometer array for the SPIRE instrument

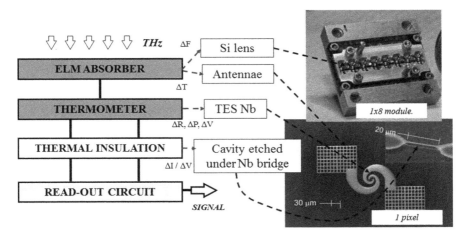

Fig. 2.18   Antenna-coupled TES Nb bolometer at NIST-VTT

A constant current bias is applied, and thus a voltage read-out is performed to ensure a stable operation with respect to the strong electro-thermal feedback induced by the very high positive TCR (typically $> 1,000\,\%/K$).

The latest publications report video-rate THz passive imaging where the camera reads-out 64 parallel TES microbolometers operated within a turn-key commercial closed-cycle cryocooler [21]. This result constitutes the first demonstration of passive THz imaging in the 0.3–0.65 THz range for any of the detector technologies considered.

As a general comment, monolithic bolometers that consist of a planar antenna integrated with a matching network and a thermometer are currently the only

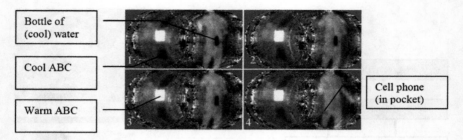

Bottle of
(cool) water

Cool ABC

Warm ABC

Cell phone
(in pocket)

Fig. 2.19 4 frames extracted from a 6 Hz video sequence (NIST-VTT)

Fig. 2.20 The PACS 32 × 64 pixel focal plane developed at CEA for the ESA Herschel project

practical solution for the THz region (>300 GHz). Such integrated sensors are easier to manufacture, more reliable, smaller, lighter and much less expensive than
waveguide optical coupling devices like feed horns. The integration also allows the
use of linear or two-dimensional arrays for imaging applications without a dramatic
increase in cost and weight of the system.

In return, specific care has to be undertaken to handle the optical coupling losses
that are encountered because of the small size and small effective collecting aperture
of the integrated antennas. In particular, for antenna-coupled bolometers, one has to
deal with the compromise[9] between antenna impedance matching with the incoming
radiation and the thermal conductance that is to be minimised for high sensitivity.

One possible approach that can help in the development of integrated bolometer arrays is the use of full silicon wafer technology that is mature and offers the
possibility to achieve monolithic arrays integrated with conventional Si electronics.

On the base of such silicon technology, two bolometer focal planes arrays (FPAs)
have been designed and manufactured by CEA-Leti in collaboration with CEA-Sap
(France) for the PACS (Photodetector Array Camera and Spectrometer) photometer
[22–24]. This device is one of the instruments of the European Space Agency 'Herschel Space Observatory' that has been launched by Ariane V in 2009 (Fig. 2.20).

These two FPAs made of 32 × 64 and 16 × 32 pixels covering respectively the
60–90 and 90–130 μm bands are composed of large-format "CCD-like" buttable
18 × 16 pixels arrays (Fig. 2.21).

---

[9] Governed by the Wiedemann-Franz law.

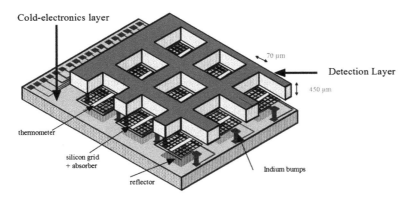

Fig. 2.21 A schematic drawing of one of the PACS 18 × 16 bolometer arrays

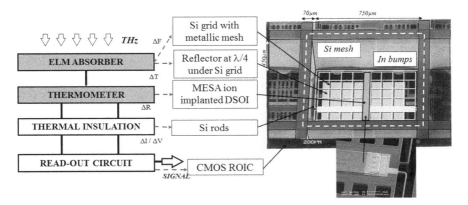

Fig. 2.22 A functional analysis of the PACS silicon bolometer

The individual pixels are defined by a $750 \times 750\,\mu m$ square aperture in a $450\,\mu m$ thick Si layer where a thin Si mesh holds a thermometer and a metallic absorbing layer. Thermometers consist of a double Silicon on Insulator (SOI) mesa where ion implantation and diffusion has been applied to achieve a very high impedance. The metallic mesh supporting the thermometer is placed above a metallic reflector deposited on the substrate layer that integrates a CMOS read-out circuit. The height of the Si grid is adjusted with respect to the reflector via indium bumps to achieve an equivalent quarter-wavelength optical absorbing cavity.

Very thin and long silicon rods connect mechanically and electrically the Si grids to the $450\,\mu m$ Si layer, whereas indium bumps ensure electrical vias towards the CMOS read-out circuit (ROIC). Like the complete monolithic bolometer array, CMOS circuit operates at 300 mK to provide impedance adaptation and multiplexing (16 to 1) reducing the number of required outputs.

This fairly large (> 1,000 pixels) focal planes based on silicon filled arrays is now operated by ESA at the L2 Lagrange spot and provides in combination to the SPIRE instrument very powerful and effective observing modes for space observations and study.

As shown by the reviewed examples, the field of cooled bolometers is quite mature and the performance provides very sensitive THz detection. However, its development is almost always carried out in laboratories or research centers, and most of the time for observation of far-infrared astronomy operated from low-background platforms or satellites. The extent of use of these high performance superconductor detectors is hindered by the available fabrication techniques and the challenging provision of many channels of read-out electronics in a low-temperature environment. In addition, cryogenic operation, especially at liquid-helium temperature, is a feature that significantly impacts the cost, the operation yield and the size for any technology. That is why, though uncooled detectors—bolometers included—are much less sensitive, they are in a position where one can think about large-scale experiments and commercial markets.

2.3 Uncooled THz Bolometers

Uncooled bolometers arrays constitute one THz technology that potentially satisfies the criteria for large volume market applications—the so-named THz "killer" application: low cost, reduced power consumption and weight, easy-to-use and high reliability, real-time 2D imaging. Indeed they combine several assets: moderate cost and high yield in production especially for the ones making use of only standard silicon (Si) processes, 2D array configuration that provides potentially large fields of view and limits the need of raster scanning, low maintenance cost related to room temperature operation, compactness because of its monolithic feature, real-time acquisition rates when pixels are processed above a CMOS read out integrated circuit (ROIC). The drawback lies in the moderate sensitivity of such incoherent uncooled sensors.

Initially developed at Honeywell [25], this technology is nowadays mature and dominates the majority of commercial and military infrared (IR) imaging applications, such as thermography, night vision (military, commercial and automotive), reconnaissance, surveillance, fire fighting, medical imaging, predictive maintenance and industrial non-destructive control.

This section presents first demonstrations of real-time THz 2D imaging with this initially IR-dedicated technology and its progressive customisation to achieve improved sensitivity in the THz range.

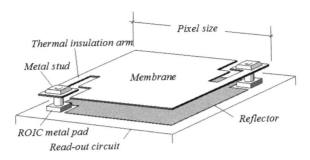

Fig. 2.23 A schematic drawing of a micro-bolometer pixel

2.3.1 Standard Thermal Infrared Bolometers Applied to THz Sensing

Figure. 2.23 shows a schematic drawing of a microbolometer pixel [4, 26]. It relies on a micro-bridge structure where a membrane is suspended above the substrate—the heat sink- by means of thermal insulation arms and metal studs. This membrane supports at the same time the absorber—e.g. TiN or NiCr- and the thermometer made of a thin film—amorphous Si or VO_x in general. This leads to a monolithic arrangement of pixels built above a substrate that usually consists of a full custom CMOS read-out circuit.

This presented overall architecture relies on a single level micro-bridge building but alternatively some bolometer designs, named "umbrella", present a two-storied structure. The first 'floor' is composed of the insulation arms and in some cases of the thermometer layer, whereas the upper level supports the absorbing layer. Such umbrella designs are especially relevant for small pixel sizes in order to overcome the growing ratio of surface occupied by the legs with respect to the sensitive area that deteriorated the optical fill factor.

To obtain a small thermal conduction between the bolometer and its surroundings, the bolometer legs are long, have a small cross-sectional area and consist of materials with a low thermal conductivity. In addition to the thermal insulation function, these legs have to provide electrical contact between the bolometer resistance and the ROIC upper metal pads. Thermal resistances as high as 50100.10^6 K/W can be reached.

For most of the commercialised IR bolometer arrays, a reflective metallic film is deposited on the surface of the underlying substrate—the CMOS ROIC—and the micro-bridge is built at a quarter-wavelength distance from this reflector (refer to Fig. 2.24). This vacuum gap is realised with surface micromachining techniques that are standard in microelectromechanical (MEMs) devices: the bolometer membrane is built on a sacrificial layer (for example polyimide) that is removed at the end of the fabrication.

The bolometer membrane constitutes a thin absorbing film with a layer sheet resistance R_f expressed in Ohms/square. If one assumes a perfect metallic reflector

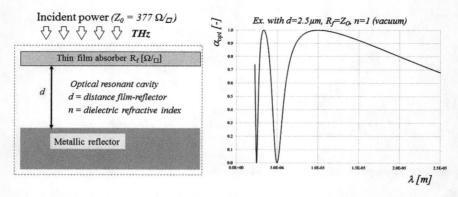

Fig. 2.24 The principle of the resonant cavity used to optimise the incoming optical absorption

placed at a distance below the membrane and the standing electromagnetic waves in the cavity filled with dielectric materials of refractive index n, the resulting optical absorption of this resonant cavity [27, 28] can be modelled as:

$$\alpha_{\text{opt}} = \frac{4y}{\left[(y+1)^2 + n^2 cot^2 (2\pi nd/\lambda)\right]}$$

with $y = (4Z_0)/R_f$.

So the bolometer resonant optical cavity has to be optimised for the targeted wavelength by the adjustment of d. For the majority of infrared imaging applications, the bolometers are optimised to detect radiation in the 8–14 μm wavelength region, the so-called long-wave infrared (LWIR) range: the corresponding vacuum gap is typically tuned from 2 to 2.5 μm. As shown in the "ideal" case of the previous figure, a high fraction of the incident infrared radiation at specific wavelengths can be absorbed due to this optical cavity.

This simple model introduces the fact that at THz frequencies, the optical absorption of this bolometer structure is drastically degraded, typically by a factor 100 at 3 THz. Moreover, commercial IR bolometer pixel pitches tend to shrink −17 μm at present time [29]—and then consequently the FPA sensitive surface is smaller than the Airy disk[10] at THz frequencies. Nevertheless, THz absorption of uncooled standard IR bolometer arrays is still sufficient to perform real-time imaging as demonstrated by several research institutes.

The first published work [30], issued from the MIT group, describes real-time THz imaging technology, using a commercial BAE 160 × 120 46.5 μm pitch pixel uncooled IR camera and a 2.52 THz (118.8 μm) methanol vapour laser pumped with a CO_2 laser ($\sim$10 mW output power). A simple experimental set-up (refer to

---

[10] The diffraction pattern resulting from a uniformly illuminated circular aperture has a bright region in the center, known as the Airy disk. Owing to this diffraction effect, the size of the Airy disk is the smallest point to which a lens or mirror can focus a beam of light.

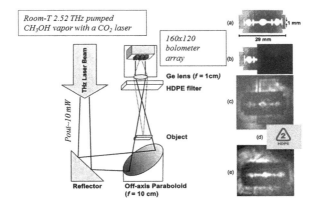

Fig. 2.25 The first published demonstration of real-time 2D imaging with commercial LWIR bolometer arrays (MIT group [30])

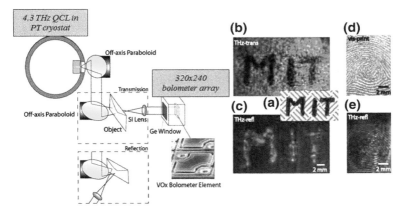

Fig. 2.26 Real-time 2D imaging with a commercial LWIR bolometer array demonstrated in transmission and reflection optical configuration (MIT group [31])

Fig. 2.25) has allowed demonstration of near-diffraction- limited real-time imaging in transmission of a razor blade concealed in a FedEx envelope.

After this feasibility demonstration, the same group used a 320×240 bolometer array to detect objects (Fig. 2.26) illuminated by a quantum cascade laser (QCL) delivering an average power of 12.5 mW at 4.3 THz [31]. Since the ambient blackbody radiation is absorbed by such sensors, the consequent undesirable common-mode LWIR signal still remaining after filtering was removed by subtraction of two different frames, acquired with THz source switched on and off. The measured NEP was ∼320 pW.

Since 1992, CEA-LETI has been involved in amorphous silicon (a-Si) uncooled microbolometer development [33, 34]. This technology has given birth in 2002 to ULIS, a French spin-off company that is now one of the world leaders in the thermal

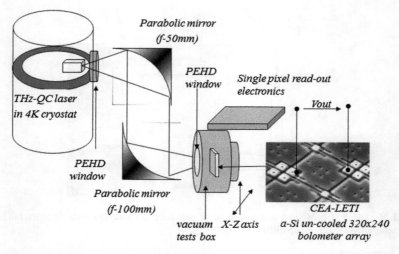

Fig. 2.27 The scheme of the CEA-Leti experimental set-up [32]

infrared sensor market [35]. In order to assess the THz performance of these 'home-made' sensors, CEA-Leti also tested the imaging of the beam delivered by a THz QCL [32].

Tests at CEA-LETI of 320 × 240 – 25 μm pitch FPAs designed and processed in their own laboratories & clean rooms have given access to optimised measurement conditions where radiometric and bolometer electro–thermal parameters (R_b, R_{th}, C_{th}...) were well-known. In particular, a PEHD window has been sealed on the vacuum package and specific prototypes enabled to address pixels individually. This configuration gave direct access to the micro-bolometer characteristics independently from the CMOS ROIC contributions to the detection chain. In return, for these tests a raster scanning has been necessary to image the full shape of the Gaussian QCL beam.

The beam profile of a 2.7 THz QCL has been acquired while a total power incident of the order of 1 mW was incident on the FPA plane. These experimental tests have been compared to 3D Finite-Element Method (FEM) simulations of the absorption of standard IR bolometric pixels in the THz domain. These first studies have generated a simulated 0.3 % absorption ratio compared to the 0.16–0.17 % extracted from the experimental data. This correspondence has been considered as satisfactory with regard to the assumptions that have been made, both on the simulation side (optical index of the stack layers) and on the experimental side (power calibration, thermal drift compensations ...).

These above results are encouraging with regard to customised bolometers since simulations of the forthcoming prototypes anticipate significantly higher sensitivity than standard IR FPAs. As a first approach, this optimisation can be done with moderate modification of the bolometer pixel, optimising the equivalent impedance in the THz domain whilst keeping the pixel dimensions unchanged.

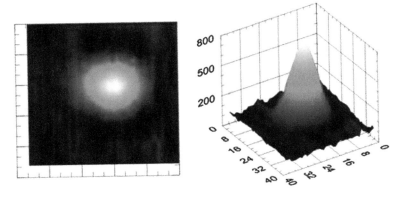

Fig. 2.28 An image of the THz QCL beam [32]

2.3.2 Bolometers with Modified Resistance of the Absorption Metallic Layer

In order to optimise the microbolometer sensitivity to THz detection, the most imme-
diate change of design that can be applied consists in depositing a metallic thin film
on the top of the suspended membrane.

As illustrated by the simple optical cavity example of the previous section, the
THz absorption can be optimised by adjusting the sheet resistance of the absorber
layer. For example, keeping the same pixel vertical dimensions (this model does not
consider horizontal effects), a sheet resistance $\sim 100\,\Omega/\square$ of is expected to produce
an absorption efficiency close to 10 % at 3 THz (refer to Fig. 2.29).

This optimisation method has been applied for the first time by NEC [36]: the
equivalent bolometer impedance of their bolometer seen from by the incoming radiation has been tuned via the addition of a thin metallic film on the upper layer of

Fig. 2.29 The modelled
absorption of the optical cavity
versus the bolometer absorber
sheet resistance at several
wavelengths 10 (IR), 100 and
300 μm ($d = 2.5$μm, $n = 1$)

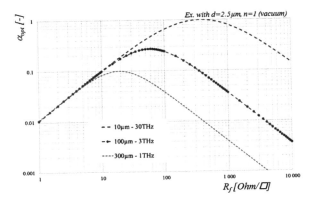

Fig. 2.30 The experimental set-up and real-time THz image of the 3.1 THz QCL beam [31]

Fig. 2.31 NEC product of handy THz camera (by courtesy of N. Oda, NEC)

a two-storied micro-bridge. Real-time images of THz beam profiles from QCLs (3.1 THz, 97 μm, average power on the FPA of 16 μW) have been obtained at 60 Hz with modified 320 × 240 THz-FPA (pixel pitch of 23.5 μm). Considering the signal-to-noise ratio of the intensity profile and transmission values of both the metal mesh filter and the vacuum package window (refer to Fig. 2.30), the NEP is estimated to be 41 ± 4 pW at 3.1 THz.

These results have motivated the commercialisation of the first real-time handy THz bolometer camera (Fig. 2.31) by NEC that integrates 320 × 240 pixels with a $NEP > 100$ pW, the complete imaging chain being considered [37].

INO (Canada) also performs R&D activities to customise the IR bolometer arrays for THz real-time imaging. Similarly to NEC, a metallic absorber film with optimum thickness has been used to maximise the THz absorption. NEP of 70 pW has been characterised at 3 THz [38].

Further, significant improvements in sensitivity can be obtained by tuning the pixel dimensions to the wavelength.

2.3.3 THz Antenna-Coupled Bolometers

A specific bolometer design for THz sensing consists in correctly placing the reflecting λ/4 backplane.

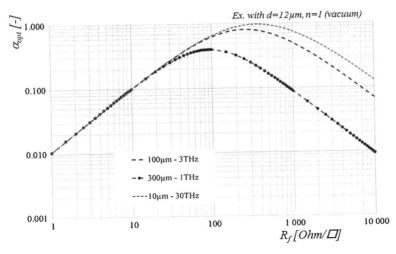

Fig. 2.32 The modelled absorption of the optical cavity versus the bolometer absorber sheet resistance at several wavelengths $10\,\mu$m (IR), 100 & $300\,\mu$m ($d = 12\,\mu$m, $n = 1$)

Figure. 2.32 illustrates the enhancement of the optical absorption versus the sheet resistance when the vacuum gap is set at $12\,\mu$m. Compared to the standard LWIR $2–2.5\,\mu$m cavity height, the NEP can be improved by another factor of 3–10 in the THz regime.

Such development has been initiated at CEA-LETI a few years ago with the constraints of remaining as close as possible to the existing amorphous silicon bolometer stack and of defining structures with a technological flow chart completely compatible with standard silicon microelectronic equipment.

In a first step, a specific 3D design was tailored for optimised imaging in the 1–3 THz range, where complementary to an optimised quarter-wavelength cavity, antennas are coupled to the bolometer to perform the optical radiation absorption [39]. This association provides separation between the two main functions of such sensors, i.e. the electromagnetic absorption and the thermometer.

The temperature transducer still consists of a suspended micro-bridge structure derived from standard IR technology whereas the optical collection function is performed by the metallic antennas and a matched quarter-wave resonant cavity processed below the antenna layer. The antenna size and shapes can be chosen independently from the bolometric device to match the illumination characteristics, in particular the frequency range and the polarization. This arrangement is very versatile with respect to frequency. Indeed, in the case of a monolithic micro-bridge that integrates simultaneously the absorber and thermometer layers, as wavelength increases, efficient optical coupling related to the geometrical fill factor implies that the suspended structures surface grows. This growth becomes quickly unpractical because of technological limitations and moreover of the large thermally isolated dimensions that result in an excessive time constant. When designing antenna-coupled

Fig. 2.33 An antenna-coupled microbolometer pixel structure (CEA-Leti)

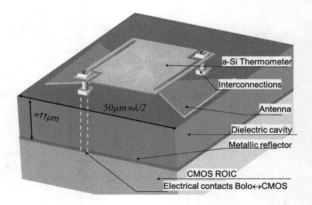

bolometers, the antennas can be sized to match the wavelength, whereas the bolometer structure can be kept small and thus the response time is preserved.

50 μm pitch pixels have been designed for an optimised sensitivity in the 1–3 THz range (Fig. 2.33). Crossed polarised antenna structures have been implemented to make the bolometer sensitive to both TE and TM polarisations, but it is possible to implement only one of these two polarisations in the pixel structure. The antenna load is integrated on the bolometer micro-bridge, thus ensuring thermal insulation from the chip substrate. In this way, the power dissipated through the Joule effect induced by THz radiation absorption can be efficiently sensed.

This complete stack is processed above a buried metallic reflector, so that a circa 11 μm thick silicon oxide layer separates the quarter-wavelength backplane and the antenna floor. The main technological key is the etching and metallisation of vias through this thick dielectric cavity that electrically connects the bolometer resistance with the CMOS upper metallic contacts. This ensures the fabrication of a monolithic 2D sensor where each pixel is read by a CMOS ROIC closely placed below. In addition, this connection going downwards to the pixel layers prevents electromagnetic perturbation of the pixels by neighboring metallic lines as encountered in [40].

The ROIC consists in a CMOS Application-Specific Integrated Circuit (ASIC) that ensures measurement of the resistances of the 320 × 240 bolometers and formats the result into a single video data stream. As in standard IR chips, the ROIC drives a row-by-row biasing and current reading in a continuous sequential electronic scan. When one row is addressed, pixels are biased and the currents flowing through the resistances are converted to voltages by Capacitance Trans-Impedance Amplifiers (CTIA) located at the end of each column. While the row N of pixels is integrated, the $N - 1$ row voltages are sampled and held, and finally multiplexed to the video output after amplification. Hence, the video output delivers an analogue flow of the multiplexed signals detected by the complete 2D array of bolometers.

As shown in Fig. 2.34, in 2010 first prototypes of 320 × 240 bolometer arrays collectively processed above a CMOS ASIC were manufactured [41].

Fig. 2.34 The first prototype of a CEA-Leti 320 × 240 antenna-coupled bolometer monolithic array

Fig. 2.35 The cartography of bolometer electrical resistances for the 63 320 × 240 bolometer arrays available on each 8" wafer (resistances are read by the CMOS ASIC)

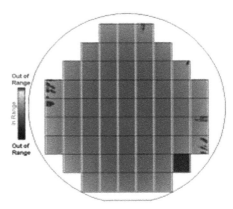

Thanks to state-of-the-art Si microelectronic facilities and robust bolometer technology know-how, a very high yield has been achieved: more than 99.5 % of the 320 × 240 pixels are functional for 56 out of 63 chips available per 200 mm wafer. An example of this statistics is visible in Fig. 2.35.

In spite of a mismatch between the bolometer resistance and the ROIC first stage that will be corrected in the next prototypes, technological yield and sensitivity have been characterised and real-time imaging has been demonstrated [42, 43].

Two pixels have been designed and fabricated with maximum spectral absorption of one polarisation tuned respectively at 1.7 and 2.4 THz. Experimental tests have confirmed the 3D FEM simulations on HFSS® that allow the plot of the optical absorption as a function of frequency [44]. The main antenna resonant frequencies are fitting the expected values and the computed and the measured data are in good correspondence, as shown in Fig. 2.36.

Figure 2.37 shows the experimental set-up used to characterise the bolometer responsivity. It has been evaluated with a quite straightforward method illuminating the FPA with a 2.4 THz Quantum Cascade Laser (QCL). The total power incoming on the FPA from the QCL is first measured with a Golay cell. Then the bolometer

Fig. 2.36 Simulated (HFSS®) simulations) and measured (in FTS) THz absorptions of the two sensors respectively designed for maximum absorptions at 1.7 and 2.4 THz [44]

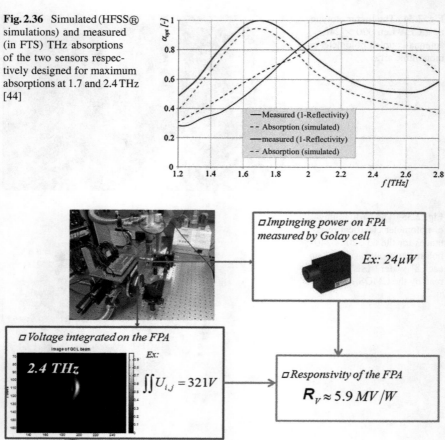

Fig. 2.37 Responsivity method applied to CEA-Leti bolometer array

array is placed on the plane to image the QCL beam. The integrated voltage signal divided by the total incident power gives the voltage responsivity with no assumptions on the antenna equivalent effective surface of the pixels; since the pixels are assembled in array, this experimental measurement takes into account the potential interferences between antenna-coupled microbolometers. The root-mean-square (rms) voltage noise is measured at the output of the complete chain including the sensor and the global noise level of the ROIC and digitizing chain. The threshold detection power is then given as $NEP = V_n/R_v$. A total noise $V_n = 400\,\mu Vrms$ has been measured. The following table reviews the test results of the two designs.

Table. 2.1 shows first that at 2.4 THz the threshold detection power is reduced between the designs at 1.7 and 2.4 THz in correspondence to the optical absorption evolution. Second, despite a non-optimised adaptation between the bolometer

Table 2.1 performances of two bolometer designs characterised at 2.4 THz

	Design 1.7 THz	Design 2.4 THz
Voltage responsivity R_V	5.9 MV/W	14.2 MV/W
Threshold detection power	68 pW	28 pW
Simulated α_{opt} at 2.4 THz	$\sim$40 %	$\sim$90 %

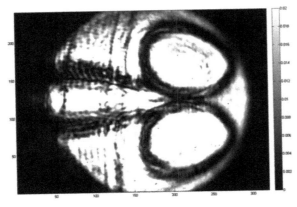

Fig. 2.38 A single frame extracted from a video sequence where scissors are translated through the THz QCL optical beam

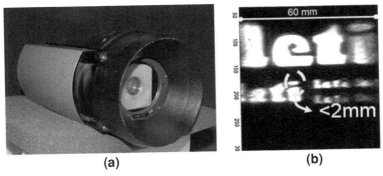

Fig. 2.39 **a** 320 × 240 antenna-coupled bolometer FPA integrated in a camera box. **b** Example of an active reflective image of a 40 × 60 mm^2 pattern with better than 2 mm spatial resolution

resistances and the CMOS reading, these results are very promising and at the top of the state-of-the-art performances of THz uncooled thermal FPAs.

Real-time 2D imaging on the full optical sensitive surface has been demonstrated with a comfortable SNR either in transmission or reflection configuration. As an example, video sequences have been acquired whilst objects like scissors hidden in a postal envelope were translated through the THz optical beam (Fig. 2.38).

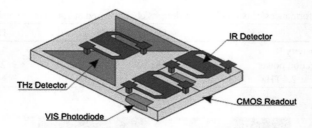

Fig. 2.40 The scheme of the 3-band monolithic sensor developed in the frame of the MUTIVIS FP7 project

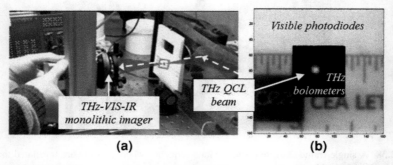

Fig. 2.41 **a** Set-up for tests of Visible and THz real-time imaging by the MUTIVIS 3-band array.
b Raw frame extracted from Visible and THz real-time video output

The 320 × 240 antenna-coupled bolometer FPA has been integrated in a camera box [(a) of Fig. 2.39] housing FPGA and front-end electronic cards that deliver video output through an Ethernet port. This compact camera has been placed at the focus of a Newton telescope that is part of a complete reflection active THz imaging demonstration system [45]. Real-time imaging of the radiation backscattered and reflected from an illuminated 40 × 60 mm^2 surface has been demonstrated with better than 2 mm spatial resolution.

As a last illustration, one has to mention the development in the MUTIVIS FP7 project of a 3-band monolithic sensor that integrates the CEA-Leti antenna-coupled pixels [46]. The visible channel pixels are based on photodiodes integrated in a CMOS read out integrated circuit (ROIC) substrate on which both IR and THz pixels have been collectively fabricated (Fig. 2.40).

Multicolor real-time video acquisition tests [47] have been performed in laboratory (Set-up photography in (a) of Fig. 2.41). As shown in Fig. 2.41, a paper optical test pattern standing in the way of a THz-Quantum Cascade Laser (QCL) beam optical path has been simultaneously imaged by visible photodiodes and THz antenna-coupled microbolometers.

The results presented in this section illustrate the technological maturity and state-of-the art sensitivity of uncooled antenna-coupled bolometer arrays. It also points out the major asset of wide focal plane arrays operating in real time at ambient temperature for THz imaging that is needed for many of the expected applications.

2.4 Conclusion

This chapter has endeavoured to give first an overview of the bolometer detection principles for non-specialists, then to illustrate how such sensors are developed for THz imaging depending on the operation temperature.

For several decades, the cooled technology has been supported by ambitious spatial programs and nowadays extremely sensitive bolometer FPAs are operating in low-background platforms or satellites. Some research institutes are now trying to transfer this know-how to larger volume applications, with some promising demonstrations like the video-rate THz camera developed by the VTT-NIST collaboration.

These cooled cameras are particularly promising for passive imaging in the atmospheric windows close to 1 THz. Active imaging making use heterodyne detection with high-T_c bolometers such as hot-electron bolometers are also to be considered as candidates for future applications, since nitrogen cooling can be performed by very compact and easy-to-use commercial system.

Nevertheless, even if uncooled bolometers are much less sensitive than cooled thermal sensors, they are in a position to meet large-scale commercial markets. Thanks to mature technology, THz-customised room temperature bolometers can potentially suit many applications that require 2D sensors with low cost in production (standard Si microelectronics) and operation (no cooling), compactness (monolithic), and 2D real-time and advanced image functions (array above CMOS). As a conclusion, bolometers are offering a wide range of technologies and operation modes that can meet a large variety of THz applications.

Acknowledgments The author wants to thank his colleagues Dr Vincent Reveret and Jérôme Meilhan for their contributions in correcting and enriching this chapter. He is also very grateful for the materials and precisions given by Naoki Oda and Aartu Lukanen. He wants to express his gratitude to Dr Matteo Perenzoni and Professor Douglas Paul, whose support helped in going through the writing of this chapter. And of course, special thanks to all CEA-Leti colleagues that have contributed–and still do–to the development of bolometer arrays applied to infrared and terahertz imaging.

References

1. R. Clark Jones, J. Opt. Soc. Am. **43**, 1–10 (1953)
2. J.C. Mather, Appl. Opt. **23**, 4 (1984)
3. J.C. Mather, J. Appl. Opt. **21**(6), 1125–1129 (1982)
4. J.J. Yon et al., in *Proceedings of AMAA 2003 Conference*, ed. by J. Valldorf and W. Gessner, pp. 137–157
5. H. Kraus, Supercond. Sci. Technol. **9**, 827–842 (1996)

6. S.P. Langley, Nature **25**, 4 (1881)
7. F. Niklaus, in *Proceedings of SPIE MEMS/MOEMS*, vol. 6836, p. 68360D (2007)
8. K.D. Irwin, G.C. Hilton, C. Enss (ed.), Topics. Appl. Phys. **99**, 63–152 (2005). doi:10.1007/10933596_3
9. A.D. Semenov et al., Supercond. Sci. Technol. **15**, R1 (2002). doi:10.1088/0953-2048/15/4/201.
10. F. Sizov, Optoelectron. Rev. **18**(1), 10–36 (2010)
11. M. Aurino et al., J. Phys.: Conference Series **97** (IOP publishing, Bristol, 2008)
12. P.L. Richard et al., Appl. Phys. Lett. **54**, 283 (1989). doi:10.1063/1.101447
13. A. Kreisler, A. Gaugue, Supercond. Sci. Technol. **13**, 1235 (2000)
14. J. Clarke et al., Appl. Phys. Lett. **48**(12), 4865–4879 (1977)
15. P.L. Richards, J. Appl. Phys. **76**(1), 1–24 (1994)
16. E. Kreysa et al., in *AIP Conference Proceedings 616*, ed. by M. De Petris, M. Gervasi, pp. 262–269 (2002)
17. M.J. Myers et al., Appl. Phys. Lett. **86**, 114103 (2005)
18. J.J. Bock et al., Space Sci. Rev. **74**, 229–235 (1995)
19. P.D. Mauskopf, Appl. Optics **36**(4), 765–771 (1997)
20. E. Grossman et al., Appl. Opt. **49**(19), E106–E120 (2010)
21. A. Luukanen et al., *International Workshop on Antenna Technology (iWAT)*, (2010)
22. F. Simoens et al., Millimeter and submillimeter detectors for astronomy II, in *Proceedings of SPIE 5498*, ed. by J. Zmuidzinas, W. S. Holland, Stafford Withington, pp. 177–186 (2004)
23. N. Billot et al., Nuclear Inst. Methods Phys. Res. A **567**, 137–139 (2006)
24. V. Revéret et al., J. Low Temp. Phys. **151**(1–2), 32–39 (2007)
25. R.A. Wood, *IEEE Solid-State Sensor and Actuator Workshop*, 5th Technical Digest, 132–135 (1992)
26. P. Kruze, D.D. Skatrud, *Uncooled Infrared Imaging Arrays and Systems*, vol. 47 de Semiconductors and semimetals (Academic Press, New York, 1997)
27. P.A. Silberg, J. Opt, Soc. Am. **47**, 575–578 (1957)
28. S. Bauer, Am. J. Phys. **60**, 257–261 (1992)
29. J.L. Tissot et al., Opt. Eng. **50**, 061006 (2011)
30. A.W.M. Lee, Hu Q., Optics Lett. **30**(19), 2563–2565 (2005)
31. A.W.M. Lee et al., IEEE Photonics Technol. Lett. **18**, 1415–1417 (2006)
32. F. Simoens et al., in *Proceedings of SPIE 7485, Millimetre Wave and Terahertz Sensors and Technology II*, ed. By K.A. Krapels and N.A. Salmon, 74850M (2009)
33. J.J. Yon et al., Proc. SPIE **5783**, 432–440 (2005)
34. J.J. Yon et al., Proc. SPIE **6940** (61), 69401W (2008)
35. B. Fieque et al., Proc. SPIE **6940**, 69401X (2008)
36. N. Oda, C. R. Phys. **11**, 496–509 (2010). doi:10.1016/j.crhy.2010.05.001
37. N. Oda et al., *Infrared Technology and Applications XXXVII*, ed. by B.F. Andresen, G.F. Fulop, P.R. Norton, in *Proceedings of SPIE* vol. 8012, pp. 80121B–80121B-9 (2011)
38. M. Bolduc et al., in *Proceedings of SPIESPIE 8023, Terahertz Physics, Devices, and Systems V: Advance Applications in Industry and, Defense*, vol. 80230C (2011). doi:10.1117/12.883507
39. E. Peytavit et al., in *Proceedings of the IRMMW-THz International Conference on Infrared and Millimeter Waves - IRMMW-THz 2005*, pp. 257–258 (2005). doi:10.1109/ICIMW.2005.1572506
40. A. Luukanen et al., Proc. SPIE **5410**, 195–201 (2004)
41. F. Simoens F.et al, in conference Infrared, Millimeter and Terahertz Waves (IRMMW-THz 2010), pp. 1–2 (2010).
42. F. Simoens et al., in *Conference Infrared, Millimeter and Terahertz Waves (IRMMW-THz 2011)*, (2011)
43. J. Meilhan, et al., Proc. SPIE **8023**, 80230E–80230E-13 (2011)
44. D.T. Nguyen et al., IEEE Trans. Terahertz. Sci. Technol. **2**(3), 299–305 (2012). doi:10.1109/TTHZ.2012.2188395

45. F. Simoens et al., in *Conference Infrared, Millimeter and Terahertz Waves (IRMMW-THz 2012)*, pp. 1–2 (2012). doi:10.1109/IRMMW-THz.2012.6380151
46. M Perenzoni et al., in *Conference Infrared Millimeter and Terahertz Waves IRMMW-THz 2010*, pp. 1–2 (2010)
47. J. Meilhan et al., in *Conference Infrared, Millimeter and Terahertz Waves (IRMMW-THz 2012)*, pp. 1–2 (2012). doi:10.1109/IRMMW-THz.2012.6380203

Chapter 3
Terahertz Plasma Field Effect Transistors

W. Knap, D. Coquillat, N. Dyakonova, D. But, T. Otsuji and F. Teppe

Abstract The channel of the field effect transistor can operate as a cavity for plasma waves. For the electrons with high enough mobility, the plasma waves can propagate in field effect transistors (FETs) channel leading to resonant Terahertz (THz) emission or detection. In the low mobility case, plasma oscillations are overdamped, but a plasma density perturbation can be induced by incoming THz radiation. This perturbation can lead to efficient broadband THz detection. We present an overview of the main experimental results concerning the plasma oscillations in field effect transistors for the generation and the detection of terahertz radiations.

Keywords Field-effect transistor · Terahertz detector · Plasma oscillations · Channel instability · Broadband detection · Resonant detection

3.1 Introduction

The channel of a field effect transistor (FET) can act as a resonator for plasma waves with a typical wave velocity $s \sim 10^6$ m/s. The fundamental frequency f of this resonator depends on its dimensions and for a gate length L of a micron ($\sim 10^{-6}$ m) or less, f can reach the terahertz (THz) range, since $f \sim s/L$. The interest in using FETs for THz applications was initiated at the beginning of 1990s by the theoretical work

W. Knap (✉) · D. Coquillat · N. Dyakonova · D. But · F. Teppe
Laboratoire Charles Coulomb, Universite Montpellier 2 and CNRS, UMR 5221,
GIS-TeraLab, F-34095 Montpellier, France
e-mail: knap.wojciech@gmail.com

W. Knap
Institute of High Pressure Physics UNIPRESS PAN, 02-845 Warsaw, Poland

T. Otsuji
Research Institute of Electrical Communication, Tohoku University, Katahira,
Aoba-Ku, Sendai 980-8577, Japan

M. Perenzoni and D. J. Paul (eds.), *Physics and Applications of Terahertz Radiation*,
Springer Series in Optical Sciences 173, DOI: 10.1007/978-94-007-3837-9_3,
© Springer Science+Business Media Dordrecht 2014

of Dyakonov and Shur [1] who predicted that a steady current flow in an asymmetric FET channel can lead to instability and spontaneous generation of plasma waves. This can produce the emission of electromagnetic radiation at the plasma wave frequency. Later, it was shown [2] that the nonlinear properties of the 2D plasma in the transistor channel can also be used for detection and mixing of THz radiation. The detection may be efficient not only in the resonant case of high electron mobility, when plasma oscillation modes are excited in the channel, but also in the non-resonant case of low mobility, where plasma oscillations are overdamped. In this case incoming THz radiation leads only to a plasma density perturbation that propagates in the channel.

Both THz emission [3–6] and detection, resonant [7–9] and non-resonant [10, 11], were observed experimentally at cryogenic, as well as at room temperatures, clearly demonstrating effects related to the excitation of plasma waves. Currently, the most promising application appears to be the broadband THz detection and imaging in the non-resonant regime. However resonant THz emission and detection by excitation of plasma waves are also interesting phenomena that deserve further exploration. An overview of the main experimental results concerning the application of FETs, first for the generation and then for detection of THz radiation is presented below.

3.2 Experiments on THz Emission from Field Effect Transistors

To measure a weak emission in the THz range, one needs a highly sensitive measurement system in which the background radiation and water vapour absorption can be avoided or compensated. In this work two different experimental setups have been used.

The first instrument was a cyclotron resonance spectrometer. It was originally conceived to investigate a weak THz cyclotron resonance emission from bulk semiconductors and their heterojunctions [12, 13]. It consisted of two superconducting coils mounted in the same helium cryostat. Two samples, a source of radiation and a detector, were mounted in a copper waveguide and placed in the cryostat at the centres of corresponding coils. In this way a high sensitivity was achieved because the system was immersed in the liquid helium and perfectly screened from the background radiation. The coils were used to make spectroscopy through the cyclotron resonance phenomenon. The emitted radiation was analysed by a narrow band, magnetically tuneable InSb cyclotron detector calibrated with conventional bulk (GaAs or InSb) cyclotron resonance emitters. Figure 3.1a shows calibration results. The cyclotron radiation from bulk InSb emitter placed in magnetic field (0.4, 0.8 and 1.2 T) was analyzed by an InSb detector. The signal was registered as a function of the magnetic field in the detector coil. It showed maxima each time the cyclotron frequency of the detector corresponded to the frequency of emitted radiation: ~0.8, 1.6, and 2.4 THz respectively.

The second instrument was a Fourier transform spectrometer operating under vacuum. It was used in some cases (mainly for GaN based devices) when the emission was strong enough to overcome the 300 K background. A fast scan and/or step scan

Fig. 3.1 (**a**) Spectra of a cyclotron resonance InSb bulk emitter and (**b**) an InGaAs HEMT emitter. The curves in (**a**) correspond to different values of the emitter magnetic field: 0.4, 0.8 and 1.2 T (from left to right). In (**b**) the emission spectra for InGaAs HEMT correspond to different source drain voltages equal to 0.3, 0.6 and 0.8 V, bottom to top [3]

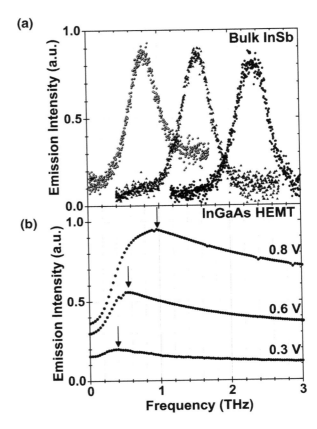

operating spectrometer equipped with an ultra-sensitive Si- bolometer was used in the 0.3–10 THz range. The emission signal was excited by square-like voltage/current pulses with frequency of a few tens of Hz and a duty cycle from 0.1–0.5. The experimental procedure was as follows: first, the reference spectrum with the unbiased transistor inside the vacuum chamber (without any applied voltage) was measured, providing information about the 300 K emission background. Then the spectra were measured with a biased transistor. To analyse only the transistor emission, the spectrometer was operating in the step scan mode and lock-in detection of the signal was synchronized to the voltage pulses. The final results were obtained by normalizing the spectra from the biased transistor by the reference spectrum.

In one of the first emission experiments, lattice-matched nanometer gate length InGaAs/InP high electron mobility transistors (HEMTs) were used. The gate length was 60 nm, and the drain–source separation was 1.3 μm. An InGaAs/InP HEMT was chosen for its high carrier mobility (resulting in high quality factor values). Typical results are shown in Fig. 3.1b. The emission spectrum was relatively broad but it displayed a maximum that shifted from ∼0.4–∼1.2 THz with increasing drain

Fig. 3.2 Emission intensity from InGaAs HEMT emitter as a function of source–drain bias in a magnetic field from 0 to 6 T [4]

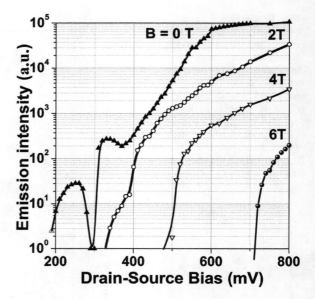

voltage. This emission appeared only above a certain threshold value of the drain voltage and sharply increased (by orders of magnitude) with increasing the bias [3]. A second superconducting coil placed at the transistor side, was used to change the conditions of the emission. The threshold voltage increased with applied magnetic field due to the increase of the channel resistance in a magnetic field [4]. The threshold behaviour for different magnetic fields is shown in Fig. 3.2.

The experimental results obtained for the InGaAs HEMT have demonstrated that THz emission appears when the drain current exceeds a certain threshold which was the argument in favour of interpreting the observed emission as due to a plasma wave instability [1]. Also, the radiation frequencies have been in reasonable agreement with estimates for fundamental plasma modes. However, some other features predicted in [1] such as a strong dependence of the radiation frequency on the gate length and the gate bias has not been observed. This is illustrated in Fig. 3.3 showing the emission spectra from InGaAs HEMTs with gate lengths varying from 50, 150 and 300 nm. One can see that the emission occurred always in the same frequency range and does not noticeably depend on the gate length.

According to the model predicted in Ref. [14] the edge instability can explain most of the observed broadband emission spectra which are not influenced by gate voltage, and gate length. This instability is related to the plasma waves propagating along the gate edges. They propagate in a direction perpendicular to the drain current, and result in broad emission spectra independent on the transistor channel dimensions [14].

Only recently, THz emission tuneable by the gate bias has been observed, in AlGaN/GaN-based high electron mobility transistors [6]. It persisted from cryogenic up to room temperature. The Fourier transform spectrometer method described above

Fig. 3.3 Spectra emission from InGaAs HEMTs with (**a**) different gate lengths: 150 and 300 nm and (**b**) different gate voltages with gate length 50 nm

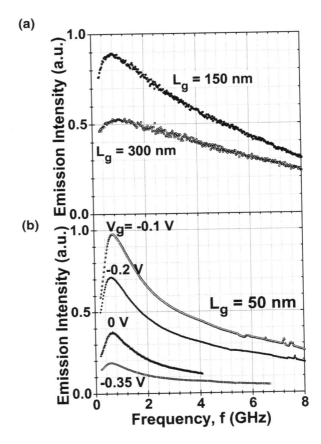

was used. The emission peak was found to be tunable by the gate voltage between 0.75 and 2.1 THz. The highest emission was obtained from transistors with a modified architecture which included the so called field plate – an additional metal layer deposited between the gate and the drain terminals. The emission appeared in a threshold manner indicating plasma instability as its origin. Figure 3.4 presents the emission spectra for one of the samples and Fig. 3.5 shows the position of the maxima of the emission as a function of the gate voltage. The dotted line illustrates the dependence of the fundamental frequency on the gate bias calculated according to the theory [1, 15].

The THz emission results obtained using the field plate GaN/AlGaN HEMTs show three main features: (i) the emission appears at a certain drain bias in a threshold-like manner (ii) the radiation frequency corresponds to the lowest fundamental plasma mode in the gated region of the transistor and (iii) the radiation frequency is tuned by the gate bias which modifies the velocity s of the plasma waves. These features are inherent attributes of the Dyakonov-Shur plasma wave instability in a gated two-

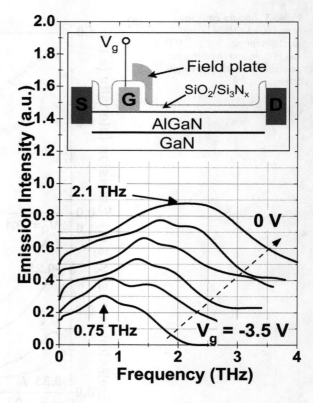

Fig. 3.4 The emission spectra for a GaN/AlGaN transistor ($L = 250\,\mathrm{nm}$) at drain voltage 4 V and different gate voltages. Inset - schematic of AlGaN/GaN HEMT with a field plate covering the gate [6]

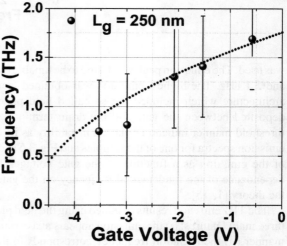

Fig. 3.5 Experimental data (*points*), and calculations (*dotted line*) of emission frequency as a function of gate bias for transistors with gate length 250 nm [6]

Fig. 3.6 Schematic of a FET
as a THz detector

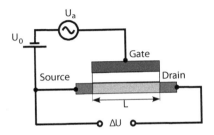

dimensional electron gas, and their presence provides a convincing argument that this phenomenon was observed.

Summarizing the current experimental state of the art concerning plasma wave instabilities and THz emission from HEMTs, one can state that different types of instabilities with threshold like behaviour of the emission intensity have been observed. Broadband gate length and gate voltage independent THz emission can be attributed to the edge instability [14]. The broadband edge emission often dominates.

Although the results on field plate HEMTs are very promising, much more experimental research is necessary to establish the architecture of the best gate-voltage-controlled THz emitters. The power levels currently registered are typically of the order of $\sim$0.1 μW. This power is relatively small when compared with other existing sources. The real importance of FET based emitters comes from the fact that using modern electronics technology, many of these elements could be easily integrated into arrays to make high power sources.

3.3 Experiments on THz Detection by Field Effect Transistors

The experimental set up for detection experiments is usually designed using monochromatic sources. The radiation is generated either by an electronic source (e.g. a source based on frequency multipliers), three backward wave oscillator (BWO) sources covering the ranges from 270–500 GHz and from 0.85–1.09 THz, or a CO_2-pumped molecular THz laser [3, 9, 10, 16, 17]. THz radiation is guided using mirrors or light pipes and coupled to the FET by contact pads and bonding wires or specially designed antennas. The experiments with polarized radiation usually show a well-defined preferential orientation of the electric field of the incident radiation related to the geometry of the antennas (or that of the bonding wires playing the role of antennas) [18]. Figure 3.6 presents the schematics of a FET as a THz detector, where the incoming radiation creates an ac voltage with amplitude U_a only between the source and gate. U_0 is the static gate swing voltage ($U_0 = V_g - V_{th}$, where U_g is voltage between source and gate parts and V_{th} is the threshold voltage of the transistor).

Typical results of detection experiments are shown in Fig. 3.7. A photovoltaic signal between the source and drain is recorded as a function of the gate voltage, i.e.,

Fig. 3.7 Photoresponse of
the GaAs/AlGaAs 0.15 μm
FET to 600 GHz radiation.
The radiation-induced source-
drain voltage is shown as a
function of the gate voltage
for different temperatures.
The arrow marks the feature
corresponding to the resonant
detection observed at the
lowest temperature [16]

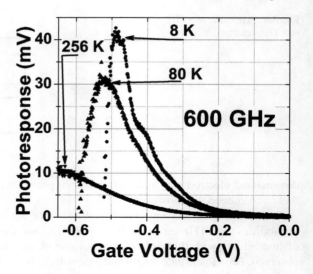

versus the carrier density in the channel. For a high carrier density (open channel) the
signal is relatively small. The signal increases and saturates when the gate voltage
approaches the threshold and rapidly decreases in the sub-threshold region [16].

Generally speaking a mechanism of the rectification is based on simultaneous
modulation of the electron concentration in the channel and the carrier drift velocity.
Because of this, in the expression for the electric current $j = env$, both the concentration n, and the drift velocity v, will be modulated at the radiation frequency and
the quadratic term appears (square detection). As a result, a dc current will appear:
$j_{dc} = e<n_1(t)v_1(t)>$, where $n_1(t)$ and $v_1(t)$ are the modulated components of n
and v, and the angular brackets denote averaging over the oscillation period $2\pi/\omega$.
Under open circuit conditions a compensating dc electric field will arise, resulting
in the photo induced source-drain voltage ΔU, see Fig. 3.6.

Depending on the frequency ω, one can distinguish two regimes of operation,
and each of them can be further divided into two sub-regimes depending on the gate
length L [11].

3.3.1 High Frequency Regime

The high frequency regime occurs when $\omega\tau > 1$, where τ is the electron momentum
relaxation time, determining the conductivity in the channel $\sigma = ne^2\tau/m$ (where n
is 2D electron concentration, e is electron charge, m is effective mass of electron).
The plasma waves have a velocity $s = (eU_0/m)^{1/2}$ and a damping time τ. Thus their
propagation distance is $L = s\tau$.

Fig. 3.8 Photoresponse versus gate voltage: (**a**) experiment and (**b**) theory, with evolution of the photoresponse with temperature (300, 180, and 8 K, bottom to top). Inset in (**a**) shows the resonant signal after subtraction of the background [9]

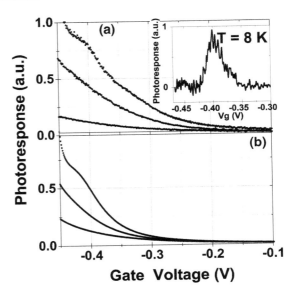

1.a. In the case of a short gate, $L < s\tau$. The plasma wave reaches the drain side of the channel, is reflected, and forms a standing wave with enhanced amplitude, so that the channel serves as a high-quality resonator for plasma oscillations. This is the case of **resonant detection**. The fundamental mode has the frequency $\sim s/L$, with a numerical coefficient depending on the boundary conditions.

In Fig. 3.8 one can clearly see the manifestation of the resonant detection: an additional peak appears with decreasing temperature. In the majority of experiments, the incoming radiation is a monochromatic beam and the source-drain voltage is recorded versus the gate voltage. The gate voltage controls the carrier density in the channel and tunes the resonant plasma frequency. A resonant enhancement of the registered voltage is observed once the resonant plasma frequency coincides with the frequency of the incoming THz radiation and the quality factor condition $Q = \omega\tau > 1$ is fulfilled. The resonance appears at low temperatures because at these temperatures the relaxation time becomes long enough (lower scattering rate). In Fig. 3.8 the results of the experiments are compared with calculations according to the theory of Ref. [2]:

$$\Delta U = \frac{1}{4}\frac{U_a^2}{U_0}f(\omega), \tag{3.1}$$

where $f(\omega) = 1 + \beta - \dfrac{1+\beta\cos(2k_0'L)}{\sinh^2(k_0''L)+\cos^2(k_0'L)}$,

$$\beta = \frac{2\omega\tau}{\sqrt{1+(\omega\tau)^2}},\ k_0' = \frac{\omega}{s}\sqrt{\frac{1}{\beta}+\frac{1}{2}},\ k_0'' = \frac{\omega}{s}\sqrt{\frac{1}{\beta}-\frac{1}{2}}.$$

1.b. Long gate, $L \gg s\,\tau$. The plasma waves excited at the source will decay before reaching the drain, so that the ac current will exist only in a small part of the channel adjacent to the source. In this case the photo induced voltage has been derived in Ref. [2] as:

$$\Delta U = \frac{U_a^2}{4U_0}(1 + \frac{2\omega\tau}{\sqrt{1 + (\omega\tau)^2}}) \qquad (3.2)$$

It is interesting to note that this formula stays valid also in the low frequency regime.

The higher quality factor condition (Q from 2–4) were obtained in high carrier mobility InGaAs transistors [9] where well defined resonances were measured. An example of resonant detection is shown in Fig. 3.9a. In Fig. 3.9b, the position of the resonant maximum is shown. As the excitation frequency increases from 1.8–3.1 THz, the plasma resonance moves to higher U_0 in an approximate agreement with theoretical predictions (dotted line).

The resonances observed in all experiments were broader than theoretically expected. A significant plasma resonance broadening appears to be one of the main unresolved problems of the resonant THz detection by FETs. The main motivation behind changing the transistor material system from GaAs/GaAlAs, as used in first the experiments [9], to InGaAs/InP [10] and using higher excitation frequencies (up to 3 THz instead of 0.6 THz) was to improve the quality factor of the response resonance. However, the experimentally observed plasma resonances remained broad.

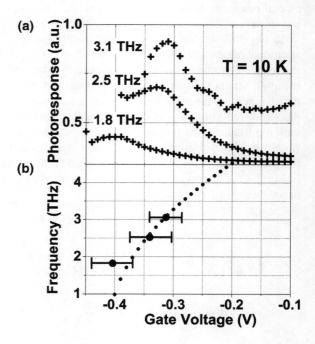

Fig. 3.9 a Photoresponse of high mobility InGaAs/InAlAs transistors at 1.8, 2.5 and 3.1 THz registered at 10 K. **b** Position of the signal maximum versus the gate voltage. Dotted line is a result of calculations [10]

Even at 3 THz the experimental quality factor Q was in the range from 2–4, while the theory predicted the quality factor should be higher by at least one order of magnitude. Understanding the origin of the broadening and minimizing it is one of the most important current experimental and theoretical challenges.

The two main hypotheses concerning broadening are: (i) the existence of oblique plasma modes [14] and (ii) an additional damping due to the leakage of gated plasmons to ungated parts of the transistor channel [19]. The first hypothesis [14] considers the devices with the gate width much greater than the gate length. In such devices, plasma waves can propagate not only in the source-drain direction but also in oblique directions. The frequency of oblique modes depends on the direction of the propagation vector and the spectrum of plasma waves is continuous. The second hypothesis, a leakage of gated plasmons to ungated parts of the channel [19], considers the fact that in the investigated HEMTs the gate covers only a small part of the source-drain distance. Therefore, the plasma under the gate cannot be treated independently of the plasma in the ungated parts. An interaction between the two plasma regions can lead not only to a modification of the resonant frequency [15], but also to a mode leakage followed by a shortening of the gated plasmon lifetimes and hence in line broadening. The relative importance of these two broadening mechanisms is still a matter of discussion.

One may decrease the role of oblique modes, by changing the geometry of the channel. Transistors with narrow channels were investigated in Refs. [20, 21].

In Fig. 3.10 present the photoresponse of a narrow multichannel transistor at 540 GHz. One can clearly see the narrowing of the resonant line at 0 mV with respect to previous means. This result clearly shows, that using FETs with narrow channels is a good way to improve the quality factor. Quality factors of the order of 10 were obtained—see Fig. 3.10b.

Another way to decrease the broadening of the plasma resonances is to apply a drain current. The drain current affects the plasma relaxation rate by driving the two-dimensional plasma in the transistor channel towards the Dyakonov-Shur plasma wave instability. As shown in Ref. [22], the resonant response in the presence of a drain current has an effective width $l = 1/\tau_{eff}$. l depends on the carrier drift velocity v according to the equation $l = 1/\tau - \gamma$, where γ is the instability increment [21, 23], which depends on the drift velocity v. For $v << s$, $\gamma = v/L$. Here L is the channel length. With increasing current, the electron drift velocity increases, leading to the increase of the quality factor $Q = \omega\tau_{eff}$. The effect of the current-driven line narrowing is illustrated in Fig. 3.10. The upper curve was measured in the current-driven detection regime with the applied drain-to-source voltage $V_{ds} = 100$ mV. A comparison of other curves ($V_{ds} = 50$ and 0 mV) shows that the line width clearly decreases with an increasing applied current. Systematic studies have shown that the quality factor could thus be increased by more than an order of magnitude [20, 21]. The good agreement between the experimental data and calculations (dotted line in Fig. 3.10) confirms the model and clearly indicates that a dc current may be used to decrease the broadening of the plasma resonances.

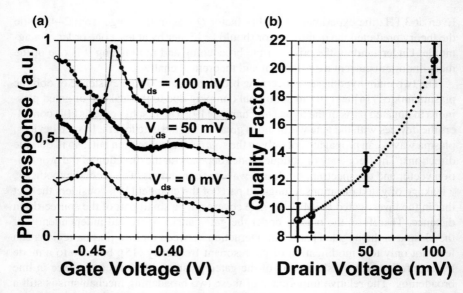

Fig. 3.10 a Photoresponse at 10 K as a function of the gate voltage. Curves were obtained at 540 GHz with a multi-channel InGaAs HEMT for different drain-source voltages V_{ds} from 0–100 mV. Curves are vertically shifted. **b** Quality factor of the resonance as a function of the source-drain voltage [20, 21]

3.3.2 Low Frequency Regime

In the low frequency regime, $\omega\tau \ll 1$. Now, the plasma waves cannot exist because of overdamping.

2a. Short gate, means that the ac current goes through the gate-to-channel capacitance practically uniformly across the whole length of the gate. This is the so-called "resistive mixer" regime [24]. For the THz frequencies this regime can apply only for transistors with extremely short gates.

2b. Long gate, means that the induced ac current will leak to the gate at a small distance l. This characteristic length for the decay of the ac voltage (current) away from the source can be written as $l = s(\tau/\omega)^{1/2}$. The decay of the THz radiation induced dc voltage along the channel is described by

$$\Delta U(x) = \frac{U_a^2}{4U_0}(1 - \exp(-2x/l)), \qquad (3.3)$$

where x is the position along the channel from the source ($x = 0$) to the drain ($x = L$).

Let now make some estimations of the characteristic length for the different cases presented above. For $\tau = 30$ fs ($\mu = 300$ cm^2/(V·s) in a Si MOSFET) and $s = 10^8$ cm/s, the high frequency regime will be achieved for radiation frequencies f greater than 5 THz; the regime $1a$ corresponds to $l < 30$ nm. For $f = 0.5$ THz (low frequency regime), one finds the characteristic gate length distinguishing regimes $2a$ and $2b$ to be around 100 nm. If the conditions of the case $1a$ are satisfied, the photoresponse will be resonant, corresponding to the excitation of discrete plasma oscillation modes in the channel. In all the other cases ($1b$, $2a$, $2b$) the FET will operate as a broad-band/nonresonant detector.

3.3.3 Nonresonant Detection Case

In the case of a nonresonant detection, the signal should increase with the decreasing carrier density, n, as $1/n$. However, as discussed in [2, 9], this conclusion is valid only well above the threshold. As shown in Fig. 3.7, for voltages close to the threshold, the signal saturates and decreases in the sub-threshold region. The main reasons for this decrease are : (i) the leakage currents and (ii) loading effects. One has to consider also that the linear dependence of the carrier density on the gate voltage is applicable only to the open state of the transistor (i.e. well above the threshold) but is not valid in the sub-threshold region.

To describe the nonresonant detection close to the threshold one has to apply another (nonlinear) approximation to the carrier density in the channel of the FET [16]:

$$n = \frac{C\eta k_B T}{e^2} \ln\left[1 + \exp\left(\frac{eU_0}{\eta k_B T}\right)\right],\tag{3.4}$$

where C is the gate capacitance per unit area, η is so called ideality factor of the transistor.

In this case the plasma wave velocity s is given by

$$s^2 = s_0^2\left(1 + \exp\left(-\frac{eU_0}{\eta k_B T}\right)\right)\ln\left(1 + \exp\left(\frac{eU_0}{\eta k_B T}\right)\right) \text{ with } s_0 = \sqrt{\frac{\eta k_B T}{m}}\tag{3.5}$$

The analytical expressions describing the signal versus gate voltage in the case of a realistic carrier density dependence and the gate leakage were given in Ref. [16]:

$$\Delta U = \frac{eu_a^2}{4ms^2}\left\{\frac{1}{1 + \kappa\exp\left(-\frac{eU_0}{\eta k_B T}\right)} - \frac{1}{\left(1 + \kappa\exp\left(-\frac{eU_0}{\eta k_B T}\right)\right)^2\left[sh^2 Q^* + \cos^2 Q^*\right]}\right\}\tag{3.6}$$

where the parameter related to the leakage current j_0

$$\kappa = \frac{j_0 L^2 m e}{2C\tau\eta^2 k_B^2 T^2}.\tag{3.7}$$

This dimensionless parameter is assumed to be small ($\kappa \ll 1$), and the factor $Q^* = \sqrt{\frac{\omega}{2\tau}} \frac{L}{s}$

It was shown that for a carrier density as in Eq. (3.4), in the absence of leakage currents and loading effects the photoresponse should saturate at a maximum value and stay constant in the subthreshold range

$$\Delta U = \frac{eU_a^2}{4\eta k_B T} \tag{3.8}$$

where U_a is the voltage induced between source and gate by incoming radiation and $\eta k_B T/e$) is the voltage that defines the exponential decay of the carrier density in the subthreshold range (see Eq. 3.4).

The results of calculations using Eq. (3.6) for different values of leakage currents are shown in Fig. 3.11. One can see that for a low leakage current the signal increases and reaches the same saturation value. This value is given by Eq. (3.8). For higher leakage currents, the signal does not reach the saturation value and a well-defined maximum near the threshold voltage is seen. Figure 3.11, illustrates that an increase of the gate leakage current suppresses the detector performance resulting in a bell-shaped photoresponse.

The temperature dependence of the nonresonant signal is presented in Fig. 3.12 and was also calculated in Ref. [16]. One can see that the temperature evolution of the calculated width and the amplitude of the photoresponse in Fig. 3.12 agrees qualitatively with the experimental results in Fig. 3.7.

For low frequency and long gate transistors (case 2b) a phenomenological approach was recently developed [25]. It relates the expected detector signal with the channel conductivity $\sigma(U)$. The resulting simple formula reads:

$$\Delta U = \frac{U_a^2}{4} \left[\frac{1}{\sigma} \frac{d\sigma}{dU} \right]_{U=U_0}. \tag{3.9}$$

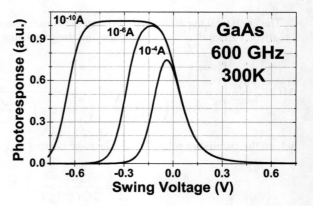

Fig. 3.11 Calculated room-temperature photoresponse of the GaAs/AlGaAs FET at 600 GHz for different gate leakage currents corresponding to 10^{-10}A, 10^{-6}A, 10^{-4}A [16]

Fig. 3.12 Calculated photoresponse of the GaAs/AlGaAs
FET at 600 GHz at four different temperatures [16]

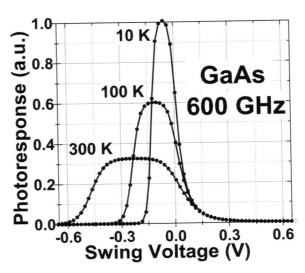

It is worth to note that Eq. (3.9) is relatively general because it takes into account the possible dependence of the mobility on gate voltage, and it remains valid even for $U_0 < 0$. This approach allows a calculation of the expected photoresponse by a simple differentiation of the transfer characteristic. This formula also describes the observed contribution of Shubnikov-de Haas oscillations of the mobility to the photoresponse: these oscillations depend on the Fermi energy, and hence on the electron concentration, which is governed by the gate voltage. The derivative in Eq. (3.9) explains the observed 90° phase difference between the photoresponse oscillations and the resistance oscillations. More detailed analysis of different detection regimes can be found in the review paper [11].

A full quantitative interpretation of the experimental results is not possible without taking into account the loading effects of the external circuit. They become important because of the exponential increase of the channel resistance below threshold. Due to loading effects, the signal below threshold will decrease even in the absence of a leakage current.

The most complete approach to the quantitative interpretation of the experimental results was recently presented in Ref. [25]. It was shown that to reproduce the experimental results one should begin by differentiation of the transfer characteristics using Eq. 3.9. Then one has to divide ΔU by a factor $(1 + R_{CH}/Z)$, where R_{CH} is the channel resistance and Z is the complex load impedance of the read-out setup. It is important that Z contains not only the load resistance of the preamplifiers but also all the parasitic capacitances. The inductance component can usually be ignored.

The experimental and theoretical results are compared in Fig. 3.13. The parasitic read-out circuit capacitances were determined by independent measurement with RLC-bridge and found to be $\sim$150 pF. The measured amplitude and phase of the signal are presented for different load resistances and modulation frequencies. One

Fig. 3.13 Amplitude (**a**) and phase (**b**) of the registered signal for different frequencies. The influence of the loading resistances on the amplitude of the signal is shown in (**c**) [26]

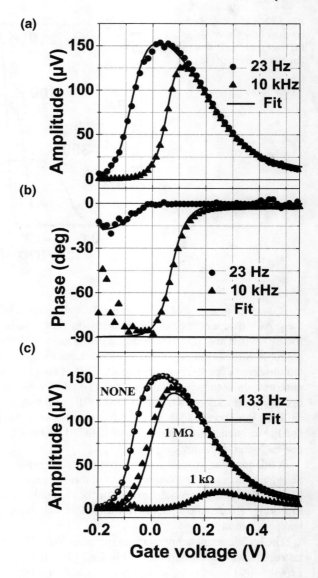

can see that by changing the modulation frequency or loading resistance we change the loading impedance and thus shift the height and position of the photoresponse maxima.

3.3.4 THz Detection by Silicon FETs

The THz detection by silicon FETs represents the most promising application of this technological domain. This is because silicon MOSFETs offer the advantages of room temperature operation, together with an easy on-chip integration with read-out electronics and high reproducibility. The first demonstration of sub-THz and THz detection by CMOS field effect transistors in silicon dates back to 2004 [27]. Tauk et al. in 2006 have shown that CMOS field effect transistors at room temperature can reach a noise equivalent power competitive with the best conventional room temperature THz detectors [28]. The first focal plane arrays in silicon technology have been designed for imaging at 300 and 600 GHz by Schuster et al. [29, 30] and Öjefors et al. [31].

Recently it has been demonstrated that by using an appropriate antenna and transistor design one can reach a record responsivity of up to a few kV/W and a NEP in the range below $10 \, pW/Hz^{0.5}$ for detectors operating in the atmospheric window around 300 GHz, which is an important result for potential applications. The record responsivity and NEP were obtained thanks to the careful design of the antenna (see Fig. 3.14).

3.3.5 Room Temperature Imaging

The room temperature imaging with FETs was demonstrated in many works. However, until now, most of the results were obtained in the sub-THz range. There are only very few results on imaging with FETs at frequencies above 1 THz. This is because in the broadband detection regime, the photovoltaic signal decreases strongly with the increase of radiation frequency either because of the reduction in coupling efficiency or because of water vapor absorption. In Fig. 3.15 images of leaves obtained with

Fig. 3.14 Responsivity as a function of frequency f for gate voltage 0.2 V. Triangles: measured points, solid line: guide for the eye. Inset: raster scan image of the source beam at 1.05 THz. ΔU is the photo-induced drain-source voltage [29]

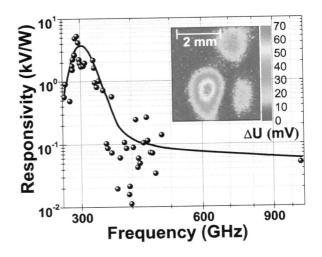

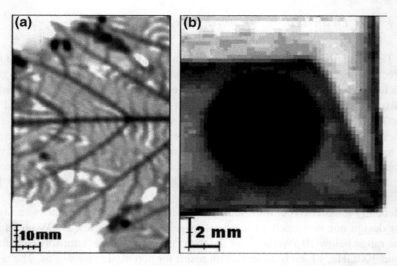

Fig. 3.15 Images of tree leaves at 300 GHz (**a**) and a medicament tablet in the envelope at 1.6 THz (**b**) obtained in the transmission mode at room temperature [17]

the 300 GHz source and a silicon MOSFET THz detector (Fig. 3.15a) are compared with an image of a medicament tablet in the envelope obtained with a CO_2 pumped far infrared laser (1.6 THz) and GaAs FET (Fig. 3.15b).

The images were recorded in transmission mode by mechanical raster scanning the sample in the X and Y directions. The good resolution, contrast, and an acquisition rate high enough for video-rate imaging systems were obtained. Generally, FETs are excellent candidates for fast room temperature THz imaging at sub-THz and THz frequencies [17].

3.3.6 Effects of the Applied Magnetic Field

Under magnetic field the dispersion law for plasma waves is written as: $\omega = [\omega_c^2 + (sk)^2]^{1/2}$, where ω_c is the cyclotron frequency (for $\omega_c = 0$ this reduces to $\omega = sk$) [2] and k is the wave vector. Thus plasma waves cannot propagate below the cyclotron frequency, since $\omega < \omega_c$ is possible only for imaginary values of the wave vector k. Therefore, in experiments with a fixed radiation frequency the photoresponse is expected to be strongly reduced when the magnetic field increases above the cyclotron resonance [32].

High magnetic field studies were performed at cryogenic temperatures on long InGaAs transistors with high mobility. These conditions correspond to the case 1*b* described in the "high frequency regime" i.e. $L >> s\tau$. The plasma waves are excited at the source and decay before reaching the drain. Oscillations analogous

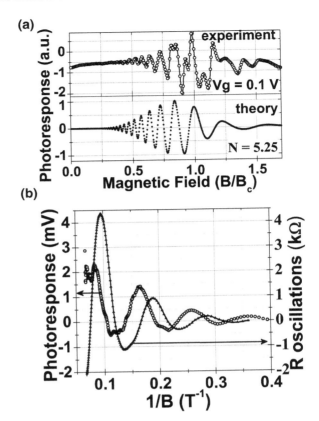

Fig. 3.16 a *Top* experimental photoresponse as a function of the magnetic field for gate voltage 0.1 V – high carrier density. *Bottom* calculations using Eq. (3.1) of Ref. [31]. **b** Comparison of the magnetoresistance and 300 GHz photoresponse oscillations

to Shubnikov-de Haas oscillations were observed. As already discussed above, the plasma waves can propagate only if the cyclotron resonance frequency is lower than the radiation frequency. In the opposite case the plasma wave vector becomes imaginary and thus the plasma oscillations are rapidly damped [32, 33].

Figure 3.16 shows the FET signal as a function of the magnetic field for a relatively low electron density. The experiments show an oscillatory character of the signal. The lower panel in Fig. 3.16a show the calculated photoresponse [33]. The theory correctly describes the oscillations of the photoresponse. Also the plasma wave damping in the post cyclotron resonance region is reproduced. This damping is probably the most striking manifestation of the importance of plasma waves in THz detection by FETs. In Fig. 3.16b the Shubnikov-de Hass oscillations of magnetoresistance are presented together with the photoresponse oscillations [34]. One can see the 90 degrees shift. This shift is due to the fact that the photoresponse is proportional to the derivative of the conductivity—see Eq. (3.9).

3.4 Detection by Double Grating Gate FETs

The problem of an effective coupling of THz radiation to a 2D electron system is of crucial importance. Metal gratings can provide a good coupling between the plasmons in the 2D electron channel and the THz radiation and were already explored in the pioneering work on plasmons in 2D electron gases [35–37]. Plasmonic THz detection by a double-grating gate field-effect transistor structure with a symmetric unit cell was studied theoretically and demonstrated experimentally. The devices consists of an interdigitated double-grating gate structure formed from two metal subgratings (G1 and G2) with different widths of fingers [38] as shown schematically in Fig. 3.17. These structures are promising for fast and sensitive detection and could be applied in future THz wireless communications. The subgratings G1 and G2 can be separately biased providing different electron densities, down to the depletion of the channel, under either grating gate [38, 39]. Structures with different ratios of the subgratings finger widths have been studied (Table of Fig. 3.18).

Each double-grating-gate FET was irradiated at the normal incidence by a THz beam at the frequency 0.24 THz. As a result, the photoresponse was observed to grow with decreasing the length of the undepleted portion of the channel compared to the structure period (under the unbiased grating-gate finger) for samples # 1–4, while it was observed to grow with increasing the length of the depleted portion of the channel compared to the structure period (under a biased grating-gate finger) for samples # 5–8 (red solid points in Fig. 3.18).

Electrodynamic modeling of the symmetric structure demonstrates that the double-grating gate produces a high THz electric field, well exceeding the electric field in the incoming THz wave, in depletion regions of the channel [40–43]. The distributions of the normal and in-plane oscillating electric-field amplitudes over the structure period shows that strong electric fields with the electric-field-enhancement factor as high as 2.5 can be excited in the channel near the edges of the longer gate finger G2 when it is biased down to the threshold voltage. Then the photoresponse for different samples was characterized by estimating the detection length in the region of the strong electric field in the depleted part of the channel. Calculated detection lengths for all eight different samples shown by the blue open points in Fig. 3.18 demonstrate a good agreement with the measured photoresponse

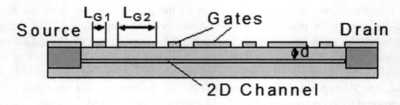

Fig. 3.17 Schematic description of the symmetric double-grating-gate device. The external THz radiation is incident normally from the top

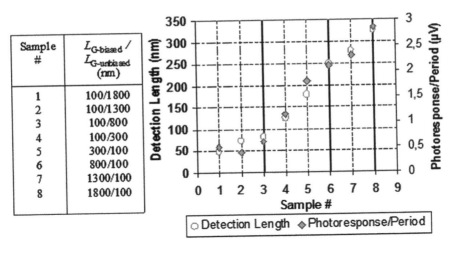

Sample #	$L_{G\text{-biased}} / L_{G\text{-unbiased}}$ (nm)
1	100/1800
2	100/1300
3	100/800
4	100/300
5	300/100
6	800/100
7	1300/100
8	1800/100

Fig. 3.18 Calculated detection length (*blue open circles*) and measured photoresponse (*red solid diamonds*) for different samples under different bias conditions. The table shows the characteristics and the biasing conditions of the samples

for all samples and it was able to explain the trend of the increasing signal observed in Fig. 3.18 [43].

For symmetric reasons, the THz photoresponse can appear only if some asymmetry is introduced in the unit cell of the double grating gate structure. This can be obtained, for example, by applying a dc source-drain current in the device channel or by introducing some systematic asymmetry. Recently, asymmetric double-grating-gate structures have been proposed. A photoresponse exceeding 30 kV/W at room temperature was predicted theoretically [44] and a record responsivity approaching 23 kV/W at 200 GHz and 6.4 kV/W at 1.5 THz were demonstrated experimentally [45, 46].

3.5 Summary

In summary both THz emission and detection, resonant and non-resonant, were observed experimentally at cryogenic, as well as at room temperatures, clearly demonstrating effects related to the excitation of plasma waves. Currently, the most promising application appears to be the broadband THz detection and imaging in the non-resonant regime. However resonant THz emission and detection by the excitation of plasma waves can also produce interesting phenomena that deserve further exploration.

Acknowledgments This work was partially supported by JST-ANR project WITH. We thank also CNRS guided GDR-I and GDR networks on Terahertz Science and Technology.

References

1. M. Dyakonov, M. Shur, Shallow water analogy for a ballistic field effect transistor: New mechanism of plasma wave generation by dc current. Phys. Rev. Lett. **71**(15), 2465–2468 (1993)
2. M.I. Dyakonov, M.S. Shur, Plasma wave electronics: novel terahertz devices using two dimensional electron fluid. IEEE Trans. Electron Devices **43**(10), 1640–1645 (1996)
3. W. Knap, J. Lusakowski, T. Parenty, S. Bollaert, A. Cappy, V.V. Popov, M.S. Shur, Terahertz emission by plasma waves in 60 nm gate high electron mobility transistors. Appl. Phys. Lett. **84**(13), 2331–2333 (2004)
4. N. Dyakonova, F. Teppe, J. Lusakowski, W. Knap, M. Levinshtein, A.P. Dmitriev, M.S. Shur, S. Bollaert, A. Cappy, Magnetic field effect on the terahertz emission from nanometer InGaAs/AlInAs high electron mobility transistors. J. Appl. Phys. **97**(11), 114313–114315 (2005)
5. N. Dyakonova, A. El Fatimy, J. Lusakowski, W. Knap, M.I. Dyakonov, M.A. Poisson, E. Morvan, S. Bollaert, A. Shchepetov, Y. Roelens, C. Gaquiere, D. Theron, A. Cappy, Room-temperature terahertz emission from nanometer field-effect transistors. Appl. Phys. Lett. **88**(14), 141903–141906 (2006)
6. A. El Fatimy, N. Dyakonova, Y. Meziani, T. Otsuji, W. Knap, S. Vandenbrouk, K. Madjour, D. Theron, C. Gaquiere, M.A. Poisson, S. Delage, P. Prystawko, C. Skierbiszewski, AlGaN/GaN high electron mobility transistors as a voltage-tunable room temperature terahertz sources. J. Appl. Phys. **107**(2), 024504–024504 (2010)
7. L. Jian-Qiang, M.S. Shur, J.L. Hesler, S. Liangquan, R. Weikle, Terahertz detector utilizing two-dimensional electronic fluid. IEEE Electron Device Lett. **19**(10), 373–375 (1998)
8. J.-Q. Lu, M.S. Shur, Terahertz detection by high-electron-mobility transistor: Enhancement by drain bias. Appl. Phys. Lett. **78**(17), 2587–2588 (2001)
9. W. Knap, Y. Deng, S. Rumyantsev, M.S. Shur, Resonant detection of subterahertz and terahertz radiation by plasma waves in submicron field-effect transistors. Appl. Phys. Lett. **81**(24), 4637–4639 (2002)
10. A. El Fatimy, F. Teppe, N. Dyakonova, W. Knap, D. Seliuta, G. Valusis, A. Shchepetov, Y. Roelens, S. Bollaert, A. Cappy, S. Rumyantsev, Resonant and voltage-tunable terahertz detection in InGaAs/InP nanometer transistors. Appl. Phys. Lett. **89**(13), 131923–131926 (2006)
11. W. Knap, M. Dyakonov, D. Coquillat, F. Teppe, N. Dyakonova, J. Lusakowski, K. Karpierz, M. Sakowicz, G. Valusis, D. Seliuta, I. Kasalynas, A.E. Fatimy, Y. Meziani, T. Otsuji, Field effect transistors for terahertz detection: physics and first imaging applications. J. Infrared Millimeter Waves **30**(12), 1319–1337 (2009). doi:10.1007/s10762-009-9564-9
12. W. Knap, D. Dur, A. Raymond, C. Meny, J. Leotin, S. Huant, B. Etienne, A far-infrared spectrometer based on cyclotron resonance emission sources. Rev. Sci. Instrum. **63**(6), 3293–3297 (1992)
13. C. Chaubet, A. Raymond, W. Knap, J.Y. Mulot, M. Baj, J.P. Andre, Pressure dependence of the cyclotron mass in n-GaAs-GaAlAs heterojunctions by FIR emission and transport experiments. Semicond. Sci. Technol. **6**(3), 160 (1991)
14. M. Dyakonov, Boundary instability of a two-dimensional electron fluid. Semiconductors **42**(8), 984–988 (2008). doi:10.1134/s1063782608080186
15. V. Ryzhii, A. Satou, W. Knap, M.S. Shur, Plasma oscillations in high-electron-mobility transistors with recessed gate. J. Appl. Phys. **99**(8), 084505–084507 (2006)
16. W. Knap, V. Kachorovskii, Y. Deng, S. Rumyantsev, J.Q. Lu, R. Gaska, M.S. Shur, G. Simin, X. Hu, M.A. Khan, C.A. Saylor, L.C. Brunel, Nonresonant detection of terahertz radiation in field effect transistors. J. Appl. Phys. **91**(11), 9346–9353 (2002)
17. S. Nadar, H. Videlier, D. Coquillat, F. Teppe, M. Sakowicz, N. Dyakonova, W. Knap, D. Seliuta, I. Kasalynas, G. Valusis, Room temperature imaging at 1.63 and 2.54 THz with field effect transistor detectors. J. Appl. Phys. **108**(5), 054505–054508 (2010)

18. M. Sakowicz, J. Lusakowski, K. Karpierz, M. Grynberg, W. Knap, W. Gwarek, Polarization sensitive detection of 100 GHz radiation by high mobility field-effect transistors. J. Appl. Phys. **104**(2), 024519 (2008). doi:10.1063/1.2957065

19. V.V. Popov, O.V. Polischuk, W. Knap, A. El Fatimy, Broadening of the plasmon resonance due to plasmon-plasmon intermode scattering in terahertz high-electron-mobility transistors. Appl. Phys. Lett. **93**(26), 263503–263503 (2008)

20. A. Shchepetov, C. Gardes, Y. Roelens, A. Cappy, S. Bollaert, S. Boubanga-Tombet, F. Teppe, D. Coquillat, S. Nadar, N. Dyakonova, H. Videlier, W. Knap, D. Seliuta, R. Vadoklis, G. Valusis, Oblique modes effect on terahertz plasma wave resonant detection in InGaAs/InAlAs multichannel transistors. Appl. Phys. Lett. **92**(24), 242103–242105 (2008)

21. S. Boubanga-Tombet, F. Teppe, D. Coquillat, S. Nadar, N. Dyakonova, H. Videlier, W. Knap, A. Shchepetov, C. Gardes, Y. Roelens, S. Bollaert, D. Seliuta, R. Vadoklis, G. Valusis, Current driven resonant plasma wave detection of terahertz radiation: Toward the Dyakonov-Shur instability. Appl. Phys. Lett. **92**(21), 212101–212103 (2008)

22. A. V. Chaplik, Possible crystallization of charge carriers in the inversion layers of low density. Sov. Phys. JETP **35**, 395 (1972), **62**, c.746–753 (1972)

23. D. Veksler, F. Teppe, A.P. Dmitriev, V.Y. Kachorovskii, W. Knap, M.S. Shur, Detection of terahertz radiation in gated two-dimensional structures governed by dc current. Phys. Rev. B **73**(12), 125328 (2006)

24. A. Lisauskas, U. Pfeiffer, E. Ojefors, P.H. Bolivar, D. Glaab, H.G. Roskos, Rational design of high-responsivity detectors of terahertz radiation based on distributed self-mixing in silicon field-effect transistors. J. App. Phys. **105**(11), 114511–114517 (2009)

25. M. Sakowicz, M.B. Lifshits, O.A. Klimenko, F. Schuster, D. Coquillat, F. Teppe, W. Knap, Terahertz responsivity of field effect transistors versus their static channel conductivity and loading effects. J. Appl. Phys. **110**(5), 054512–054516 (2011)

26. H. Videlier, S. Nadar, M. Sakowicz, T. Trinhvandam, D. Coquillat, F. Teppe, N. Dyakonova, W. Knap, T. Skotnicki, Terahertz broadband detection using silicon MOSFET (2009)

27. W. Knap, F. Teppe, Y. Meziani, N. Dyakonova, J. Lusakowski, F. Boeuf, T. Skotnicki, D. Maude, S. Rumyantsev, M.S. Shur, Plasma wave detection of sub-terahertz and terahertz radiation by silicon field-effect transistors. Appl. Phys. Lett. **85**(4), 675–677 (2004)

28. R. Tauk, F. Teppe, S. Boubanga, D. Coquillat, W. Knap, Y.M. Meziani, C. Gallon, F. Boeuf, T. Skotnicki, C. Fenouillet-Beranger, D.K. Maude, S. Rumyantsev, M.S. Shur, Plasma wave detection of terahertz radiation by silicon field effects transistors: Responsivity and noise equivalent power. Appl. Phys. Lett. **89**(25), 253511–253513 (2006)

29. F. Schuster, D. Coquillat, H. Videlier, M. Sakowicz, F. Teppe, L. Dussopt, B. Giffard, T. Skotnicki, W. Knap, Broadband terahertz imaging with highly sensitive silicon CMOS detectors. Optics Express **19**(8), 7827–7832 (2011). doi:10.1364/oe.19.007827

30. F. Schuster, W. Knap, V. Nguyen, Terahertz imaging achieved with low-cost CMOS detectors. Laser Focus World **47**(7), 37–41 (2011)

31. E. Ojefors, U.R Pfeiffer, A. Lisauskas, H.G. Roskos, A 0.65 THz focal-plane array in a quarter-micron CMOS process technology. IEEE J. Solid-State Circuits, **44**(7), 1968–1976 (2009)

32. M.B. Lifshits, M.I. Dyakonov, Photovoltaic effect in a gated two-dimensional electron gas in magnetic field. Phys. Rev. B **80**(12), 121304 (2009)

33. S. Boubanga-Tombet, M. Sakowicz, D. Coquillat, F. Teppe, W. Knap, M.I. Dyakonov, K. Karpierz, J. Lusakowski, M. Grynberg, Terahertz radiation detection by field effect transistor in magnetic field. Appl. Phys. Lett. **95**(7), 072103–072106 (2009)

34. O.A. Klimenko, Y.A. Mityagin, H. Videlier, F. Teppe, N.V. Dyakonova, C. Consejo, S. Bollaert, V.N. Murzin, W. Knap, Terahertz response of InGaAs field effect transistors in quantizing magnetic fields. Appl. Phys. Lett. **97**(2), 022111–022113 (2010)

35. N.T. Thomas, Plasmons in inversion layers. Surf. Sci. **98**(1–3), 515–532 (1980). doi:10.1016/0039-6028(80)90533-6

36. X.G. Peralta, S.J. Allen, M.C. Wanke, N.E. Harff, J.A. Simmons, M.P. Lilly, J.L. Reno, P.J. Burke, J.P. Eisenstein, Terahertz photoconductivity and plasmon modes in double-quantum-well field-effect transistors. Appl. Phys. Lett. **81**(9), 1627–1629 (2002)

37. E.A. Shaner, M. Lee, M.C. Wanke, A.D. Grine, J.L. Reno, S.J. Allen, Single-quantum-well grating-gated terahertz plasmon detectors. Appl. Phys. Lett. **87**(19), 193503–193507 (2005)
38. T. Otsuji, M. Hanabe, T. Nishimura, E. Sano, A grating-bicoupled plasma-wave photomixer with resonant-cavity enhanced structure. Opt. Express **14**(11), 4815–4825 (2006)
39. T. Otsuji, Y.M. Meziani, M. Hanabe, T. Ishibashi, T. Uno, E. Sano, Grating-bicoupled plasmon-resonant terahertz emitter fabricated with GaAs-based heterostructure material systems. Appl. Phys. Lett. **89**(26), 263502–263503 (2006)
40. V.V. Popov, M.S. Shur, G.M. Tsymbalov, D.V. Fateev, Higher-order plasmon resonances in GaN field-effect transistor arrays. Int. J. High Speed Electron. Syst. **17**, 557–566 (2007)
41. V.V. Popov, O.V. Polischuk, T.V. Teperik, X.G. Peralta, S.J. Allen, N.J.M. Horing, M.C. Wanke, Absorption of terahertz radiation by plasmon modes in a grid-gated double-quantum-well field-effect transistor. J. Appl. Phys. **94**(5), 3556–3562 (2003)
42. V.V. Popov, G.M. Tsymbalov, M.S. Shur, Plasma wave instability and amplification of terahertz radiation in field-effect-transistor arrays. J. Phys.: Condens. Matter **20**(38), 384208 (2008)
43. D. Coquillat, S. Nadar, F. Teppe, N. Dyakonova, S. Boubanga-Tombet, W. Knap, T. Nishimura, T. Otsuji, Y.M. Meziani, G.M. Tsymbalov, V.V. Popov, Room temperature detection of sub-terahertz radiation in double-grating-gate transistors. Opt. Express **18**(6), 6024–6032 (2010)
44. V.V. Popov, D.V. Fateev, T. Otsuji, Y.M. Meziani, D. Coquillat, W. Knap, Plasmonic terahertz detection by a double-grating-gate field-effect transistor structure with an asymmetric unit cell. Appl. Phys. Lett. **99**(24), 243504–243504 (2011)
45. T. Watanabe, S. Boubanga Tombet, Y. Tanimoto, D. Fateev, V. Popov, D. Coquillat, W. Knap, Y. Meziani, Y. Wang, H. Minamide, H. Ito, T. Otsuji, InP-and GaAs-based plasmonic high-electron-mobility transistors for room-temperature ultrahigh-sensitive terahertz sensing and imaging. IEEE Sensors J. **13**(1), 80–99 (2013)
46. T. Watanabe, A. Satou, T. Suemitsu, W. Knap, V. Popov, T. Otsuji, Plasmonic tera-hertz monochromatic coherent emission from an asymmetric chirped dual-grating-gate InAlAs/InGaAs/InP HEMT with highly asymmetric resonantcavities. In: *Optical Terahertz Science and Technology Conference (OTST), 2013 International*, 1–5 Apr 2013, pp. F2B–3

Part II
THz Sources

Chapter 4
Quantum Cascade Lasers

Douglas J. Paul

Abstract THz sources are the fundamental requirement for THz technology even for passive detection systems where sources are required for heterodyne detection to improve detectivity. A review is given for one such THz source, the quantum cascade laser. Quantum cascade lasers are unipolar, intersubband photon emitters that can be tuned through the width of quantum wells. Such lasers allow high output powers, up to hundreds of mWs at THz frequencies and provide narrow linewidths ideal for spectroscopic applications. The physics behind the operation of quantum cascade lasers will be reviewed before methods of optimising the laser design for THz operation will be discussed. Present THz performance will be discussed before future performance requirements will be reviewed.

Keywords Terahertz source · Quantum cascade laser · Quantum well · Intersubband transitions · Quantum tunneling · Injection

4.1 Background

The terahertz part of the electromagnetic spectrum, 300 GHz frequency (1 mm wavelength) to 10 THz (30 μm), has historically had a relative lack of high power and convenient radiation sources. Whilst a number of applications for THz technology may use passive detection systems, practical sources of THz radiation are fundamental to improved detection and to many of the applications being considered for THz radiation [1, 2]. Heterodyne detection requires a source to undertake non-linear mixing with the detected signal, potentially allowing many orders of magnitude improvement in detectivity [1]. Therefore, even so-called passive detection systems

D. J. Paul (✉)
School of Engineering, University of Glasgow, Rankine Building,
Oakfield Avenue, Glasgow G12 8LT, UK
e-mail: Douglas.Paul@glasgow.ac.uk

M. Perenzoni and D. J. Paul (eds.), *Physics and Applications of Terahertz Radiation*,
Springer Series in Optical Sciences 173, DOI: 10.1007/978-94-007-3837-9_4,
© Springer Science+Business Media Dordrecht 2014

of THz radiation can be improved by the use of THz sources. For spectroscopic measurements, narrow line width sources are essential. Therefore, stable narrow band sources are required with high power and high tunability for many of the potential THz applications.

Interband semiconductor lasers for THz wavelengths require extremely small band gaps, well below the band gaps of most readily available and cheap semiconductor materials. Indeed, the lowest frequency available interband lasers are based on lead salt materials with emission frequencies of 15 THz and above. A secondary problem is that the laser output power is proportional to the frequency of emission and since the THz part of the spectrum is about three orders of magnitude below the visible, all THz lasers with comparable designs to visible lasers will have around three orders of magnitude less output power than their visible equivalents.

Intersubband lasers, however, are much more suited to the THz part of the spectrum than inter band lasers. Quantum wells where one material with a lower conduction band is sandwiched between two barrier materials of a higher conduction band can readily be fabricated using molecular beam epitaxy (MBE) or chemical vapour deposition (CVD). Kazarinov and Suriz [3, 4] originally suggested the concept of an intersubband laser in 1971, but the first realisation of a quantum cascade laser (QCL) had to wait until 1994 when Faist et al. [5] at Bell Laboratories demonstrated the first QCL at mid-infrared frequencies. The emission frequency of the laser can be tuned by choosing the correct quantum well width where the emission frequency $\propto 1/[\text{well width}]^2$. At low temperatures and for intersubband transitions below the optical phonon energy (36 meV or 8.7 THz in GaAs), most electrons possess insufficient energy to allow LO-optical phonon scattering which allows long lifetimes which can be used to generate population inversion and gain. The QCL is an intersubband laser which uses multiple quantum wells to emit photons at THz or mid-infrared frequencies so that unlike an interband laser, a single electron can emit many photons by cascading through all or most of the quantum wells in the laser, thereby giving the laser its name. Such descriptions mask many of the difficulties in producing gain at THz frequencies and it was not until 2001 that Köhler and colleagues at the Scuola Normale Superiore in Pisa, Italy demonstrated the first THz QCL [6].

This chapter will provide an overview of the background physics of quantum wells and quantum mechanical tunnelling before describing the requirements to produce a QCL. The gain mechanisms, active regions and waveguide losses will be discussed before experimental examples will be reviewed. Finally, a number of the limitations of QCLs will be reviewed before future requirements are discussed. A number of review articles cover the design and operation of THz QCLs and should be consulted for further information [7, 8].

4.2 Quantum Wells and Quantum Mechanical Tunnelling

A QCL is produced using quantum wells and quantum mechanical tunnel barriers. Before discussing the design of the QCL, the basic physics of quantum wells and barriers will be reviewed.

Quantum wells are used for many applications in micro and optoelectronics including quantum well-infrared photodetectors, semiconductor lasers and QCLs [9]. Figure 4.1 shows a simple example of a quantum well formed in the conduction band of a heterostructure system where the conduction band edge of material 1 is lower in energy than that of material 2. Electrons in the system fall into the lowest energy part of the system which is the quantum well. Rather than forming a continuum of states and filling the quantum well like a liquid filling a glass, the electrons are quantised with the electrons only occupying discrete energy states called subbands. The energy of these subbands can be found by solving Schödinger's equation using trial wave functions. For an infinitely deep quantum well of width w along the x-direction, the wave function is split into periodic repeats known as Bloch states combined with an envelope function $\psi(x) = A \sin\left[\frac{n\pi x}{w}\right]$ where A is the amplitude of the wave function can be used to find the discrete energy solutions. The lowest three subband states are shown in Fig. 4.1 along the x-direction. Each subband state also forms a parabolic energy distribution for wavenumbers parallel to the quantum well heterointerfaces. The energy, E_n of the nth electron states which form the subbands for an infinitely deep quantum well are given by

$$E = \frac{\hbar^2 k_n^2}{2m^*} + \frac{\hbar^2 k_\parallel^2}{2m^*} = \frac{\hbar^2 \pi^2 n^2}{2m^* w^2} + \frac{\hbar^2 k_\parallel^2}{2m^*} \quad n = 1, 2, 3, \ldots \quad (4.1)$$

where $\hbar$ is Planck's constant divided by 2π, k_n is the wavenumber of the n-th subband, $k_\parallel$ is the wavenumber along the quantum well and m^* is the effective mass of the electrons in the semiconductor material 1. Since $E = h\upsilon = \frac{hc}{\lambda}$ for a frequency υ and wavelength, λ for the speed of light, c, the intersubband frequency or wavelength of emission between two different subbands can easily be calculated.

The second main piece of physics for the operation of a QCL is quantum mechanical tunnelling. Figure 4.2 shows the wave functions for incident and reflected electron wavefunctions on a barrier of height V in energy and width, b. The wavefunction of the electron decays exponentially as it travels through the barrier. Provided the

Fig. 4.1 A schematic diagram of a quantum well formed in the conduction band using two different semiconductor materials. The lowest three subband wave function states for $n = 1$, 2, and 3 are plotted

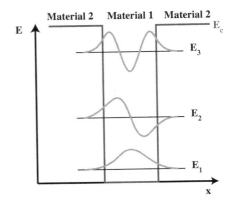

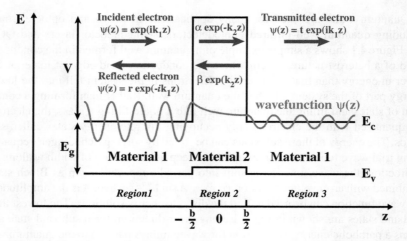

Fig. 4.2 A schematic diagram of a quantum mechanical barrier with the electron wavefunctions for incident, reflected, tunnelling and transmitted electrons shown

electron wavefunction has some finite amplitude on the other side of the barrier, the electron has a probability to tunnel through the barrier. By applying boundary conditions at the interfaces where the wavefunctions and the first derivative of the wavefunctions must be equal and continuous, the transmission coefficient can be found to be

$$T \approx \frac{16E}{V} \exp\left(-2k_2b\right) \tag{4.2}$$

for an electron of energy, E where $k_1 \left(=\sqrt{2m^*E/\hbar^2}\right)$ is the wavenumber of the electron in the conduction band of material 1 and $k_2 \left(=\sqrt{2m^*(V-E)/\hbar^2}\right)$ is the wavenumber of the electron inside the barrier. Once the transmission coefficient has been determined for the system, the current density, J through the barrier can be calculated using

$$J = q \int_0^\infty g(E)f(E)v(E)T(E)\mathrm{d}E \tag{4.3}$$

where $g(E)$ is the density of states (2D in QCLs), $f(E)$ is the Fermi function and $v(E)$ is the electron velocity.

Figure 4.3 shows the effects of placing progressively more quantum wells in close proximity with thin tunnel barriers between each of the quantum wells. Figure 4.3a shows a single quantum well with only the lowest subband state shown. In Fig. 4.3b, when a second quantum well is placed beside the first then the two lowest subbands in each of the quantum wells interact to form a set of doublet states: the lower energy subband of the double quantum well system has a symmetric wavefunction and the

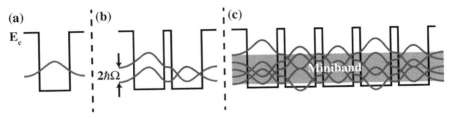

Fig. 4.3 **a** The ground subband state in a single quantum well. **b** The hybridised ground states of 2 coupled quantum wells with a symmetric-antisymmetric splitting of 2 $\hbar\Omega$. **c** A miniband formed by the coupled ground states of 5 quantum wells

higher state in energy has an antisymmetric wavefunction. There is a symmetric–antisymmetric gap due to what is termed the anti-crossing between the doublet states and this has an energy given by 2 $\hbar\Omega$ where Ω is the Rabi frequency for the transition between the two doublet states. If more quantum wells are placed together, then each of the lowest subbands overlaps with all the other quantum wells to form a miniband of extended states across all or many of the quantum wells. This is shown in Fig. 4.3c and any electrons with energy corresponding to the miniband energy can be transported from left to right if a small electric field is applied. The symmetric-antisymmetric gap and the miniband are fundamental building blocks of QCLs as will be shown in the later sections.

4.3 QCL Lasing Requirements

There are a number of reasons why the THz QCL was far harder to achieve than the mid-infrared QCL. The first is that the small transition energies of around 4–20 meV make the injection of electrons into an upper laser state and the removal of electrons from a lower laser state much more difficult by tunnelling or scattering. The population inversion required to produce gain is therefore much more difficult to achieve. Secondly, free carrier absorption from doped regions have losses which increase as the square of the wavelength and in the THz part of the spectrum can become one of the dominant loss mechanisms in the system. These losses therefore have to be managed by appropriate waveguide designs.

Considering the above comments, the requirements for a QCL to lase at THz frequencies can be summarised as:

- Subband states between which photons can be emitted at THz frequencies through intersubband transitions.
- Efficient injection of electrons into the upper laser subband state.
- Population inversion between the upper laser subband state and the lower transition state.
- Stimulated emission of photons from the intersubband transition.

- Optical cavity for feedback which will produce optical gain i.e. amplification, with a large modal overlap.
- Optical gain greater than the waveguide and other losses (e.g. free carrier losses) in the system.

4.4 Gain in Quantum Cascade Lasers

The theory behind the generation of gain from QCLs was first discussed by Kazarinov and Suris in 1972 [4] and this was later expanded by Sirtori et al. [10] to fit the experimental QCL devices. Whilst the physics can appear to get quite complicated, in essence the gain is dependent on the injection efficiency into the upper laser level, the overlap of the wavefunctions of the upper and lower laser levels and the population inversion in the system.

Figure 4.4a shows the two important regions for the design of a QCL. An injector, which may be a single subband state or a miniband of many mixed subband states with a ground wavefunction $|z_{inj}>$, injects carriers into an upper radiative subband state in the active region of the device. Carriers from this upper radiative subband with lifetime, τ_{up}, decay to a lower subband state emitting radiation with frequency, υ given by the energy difference between the two states divided by Planck's constant, h. At some electric field, F_r, the injector state $|z_{inj}>$ will anticross with the upper radiative state, $|up>$, with a splitting between the singlet and doublet states of $2\hbar\Omega$. If the intrasubband dephasing time is $\tau_\perp$ then the current density, J in the QCL as a function of the electric field, F is given by [10]

$$J = qn_s \frac{2\Omega^2 \tau_\perp}{1 + \Delta^2 \tau_\perp^2 + 4\Omega^2 \tau_{up}\tau_\perp} \tag{4.4}$$

where n_s is the sheet carrier density per period and

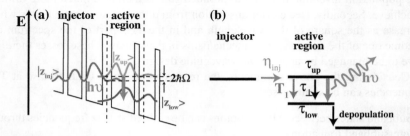

Fig. 4.4 a A schematic diagram showing the important subband states which determine the gain for an intrawell (*vertical*) radiative transition between the upper state ($|up>$) and the lower state ($|low>$). The ground state of the injector is $|inj>$. **b** A schematic diagram of the energy levels, lifetimes and scattering mechanisms for a generic QCL design

$$\Delta = \frac{E_{up} - E_{inj}}{\hbar} = \frac{q(F - F_r)}{\hbar} \left| z_{up} - z_{inj} \right| \tag{4.5}$$

$\hbar\Delta$ is the detuning from the anticrossing resonance which can also be written in terms of the centre of mass electron wavefunctions for the upper radiative state z_{up} and the injector wavefunction z_{inj}.

The ideal regime for the operation of a laser is when there is strong coupling between the injector and the upper state [10]. This corresponds to the condition $4\Omega^2 \tau_{up} \tau_\perp \gg 1$. This strong coupling regime therefore provides fast electron injection into the upper state without being limited to the tunnelling rate or any scattering rate into the upper state. Therefore, the injection into the upper state is only controlled by the upper state lifetime, τ_{up}. The maximum current flow into the upper state is then given by $J_{max} = \frac{qn_s}{2\tau_{up}}$ with only the upper state lifetime and doping density in the cascade determining the maximum current density. In the coherent picture of electron transport, electrons (holes) are transported through the tunnel barrier with Rabi oscillations at frequency Ω due to the interaction between the states $|z_{inj}>$ and $|z_{up}>$ [11].

The intrasubband dephasing time has been shown to be

$$\frac{1}{\tau_\perp} = \frac{1}{\tau_{up}} + \frac{1}{2T_1} + \frac{1}{T_2} \tag{4.6}$$

in the density matrix model [11] where T_1 is the scattering time for relaxation and T_2 is the pure dephasing contribution. Scalari et al. [8] have suggested that for transitions below the LO phonon energy where inhomogeneous broadening has been observed experimentally, $\tau_\perp$ can be estimated from the full width half maxima of the spontaneous emission linewidth, γ_{ij} using $\tau_\perp \simeq \frac{\hbar}{\gamma_{ij}}$.

The gain coefficient for QCLs can be found by forming the rate equations for the electrons moving through a three level system as described by the upper laser state, the lower laser state and the depopulation state as shown in Fig. 4.4b. For a matrix element, z_{ij}, period length, L_p and emission at wavelength λ with material with effective refractive index of n_{eff} the gain coefficient is given by [12]

$$g = \eta_{inj} \tau_{up} \left(1 - \frac{\tau_{low}}{\tau_\perp} \right) \frac{4\pi q}{\lambda \varepsilon_0 n_{eff}} \frac{z_{ij}^2}{L_p} \frac{1}{2\gamma_{ij}} \tag{4.7}$$

where η_{inj} is the injection efficiency. As maximising the gain is the main aim of an active region design, the main parameters which can be engineered by design are the matrix element and lifetimes. η_{inj} is a very difficult parameter to measure and so most designs assume unity injection efficiency and then use the lifetimes to account for any leakage currents which might reduce the value. γ_{ij} is typically set by the material and growth parameters as both non-parabolicity and interface roughness can both increase γ_{ij}.

4.5 Active Region Designs

The two most successful designs for GaAs QCLs at THz frequencies are the bound-to-continuum design [13] (Fig. 4.5a) and the phonon depopulation design [14, 15] (Fig. 4.5b). A number of other designs have also been demonstrated such as superlattice designs [6] but are now generally agreed by the QCL community to provide less gain and higher current thresholds. The bound-to-continuum design uses a miniband to both act as a fast depopulation mechanism through electron–electron scattering and then also as an injector into a single bound state. The design is aimed at optimising the injection efficiency into the upper radiative subband by only having a single bound state between the first and second minibands [8, 13]. The phonon depopulation design of Fig. 4.5b instead uses a wider quantum well where the non-radiative transition from the upper state to the lower singlet and doublet corresponds to the LO optical phonon energy [14, 16–18]. In III–V materials, the LO optical phonon scattering is a resonant process with the lifetime minimised at the LO optical phonon energy but in Group IV materials such as SiGe, the non-polar deformation potential scattering result in a non-resonant process. Therefore for SiGe designs, the depopulation

Fig. 4.5 The band diagrams of the two most successful THz QCL designs: **a** bound-to-continuum and, **b** phonon depopulation. The *red arrow* indicates the radiative transition while the *purple arrow* in (**b**) indicates the longitudinal optical phonon scattering process used for fast depopulation

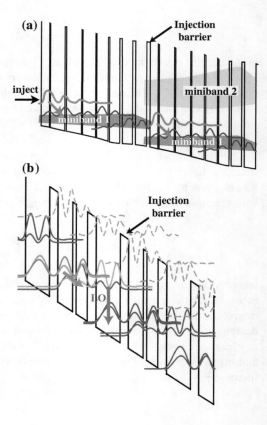

transition energy should be greater than the radiative transition energy to achieve population inversion [19, 20]. For all THz designs, especially for the lower emission energies, one of the key issues is the prevention of reabsorption. If transitions exist in the injector, depopulation region or between the radiative state and upper subbands or minibands with the same energy as the radiative transition, then the gain is significantly reduced by the reabsorption of emitted photons. It is this reabsorption that makes the design of THz QCLs especially difficult since transition energies may be comparable to miniband widths in the injector region.

4.6 Waveguide Designs and Losses

The purpose of a waveguide is to confine the radiation and to provide optical feedback. For intersubband transitions between electron subbands, the optical mode is always TM or z-polarised so that all QCLs with be edge emitting. The threshold condition for lasing is achieved when the round trip of the cavity mode is in phase with the original wave (i.e. 2π phase difference) and the gain at that frequency equals the waveguide and other loss in the system. The design of low loss waveguide at THz frequencies is predominantly occupied with managing the large optical mode but also with free carrier losses which can be large at THz frequencies. As an example, for a 3 THz emitter the wavelength in free space is $100\,\mu m$ and with the refractive index of GaAs of around 3.3, the optical mode in the waveguide will occupy a length in the order of $30\,\mu m$. To grow $30\,\mu m$ of active heterolayers is extremely time consuming and expensive, so the gain region in the vertical direction is typically $10–15\,\mu m$ deep for THz QCLs. Therefore, the key issue is optimising the modal overlap with the gain region of the device whilst keeping the waveguide losses low.

The required material gain, G, to overcome a waveguide design with total losses of α_{tot} is given by [10]

$$G = gJ_{th} = \frac{\alpha_w + \alpha_m + \alpha_{fc}}{\Gamma} = \frac{\alpha_{tot}}{\Gamma} \qquad (4.8)$$

where Γ is the modal overlap with the gain region and is defined in terms of the electric field, F in the z-direction of the wafer as

$$\Gamma = \frac{\int_{active\ region} |F(z)|^2}{\int_{\infty}^{\infty} |F(z)|^2} \qquad (4.9)$$

The total loss, α_{tot} is the sum of the (undoped) waveguide loss, α_w, the mirror losses, α_m and the free-carrier losses from the doping, α_{fc}. g is the gain coefficient as discussed above in Sect. 4.4. The free carrier absorption is related to the absorption of radiation exciting a carrier to an intermediate state where it interacts with phonons or defects in the system [21]. The free carrier absorption for N free carriers in group IV and III–V materials is given by

$$\alpha_{\text{fc}} = \frac{Nq^2\lambda^2}{4\pi^2\varepsilon_0 c^3 n m^{*2}\mu} \tag{4.10}$$

where q is the electron charge, λ is the wavelength of radiation, N is the electron density, ε_0 is the permittivity of a vacuum, c is the speed of light, n the refractive index and μ is the mobility of the material [22]. In III–V materials, an additional polar mode due to the polar optical phonons must also be added which increases the free carrier absorption with an additional $\lambda^{2.5}$ term [21].

For a ridge waveguide with facets at each end, the mirror losses are given by the equation

$$\alpha_m = \frac{1}{2L}\ln\left(R_1 R_2\right) \tag{4.11}$$

where L is the length of the waveguide and R_1 and R_2 are the reflectivities at each facet. For a cleaved or etched facet, the reflectivites will be about 0.33 since air will have $n \simeq 1$. By using facet coatings such as a high conductivity metal on a thin insulator, the reflectivity can be improved to > 0.95 to reduce loss.

There are two main types of waveguide design that will be discussed. The two designs are all aimed at confining the optical mode in the vertical direction and in all cases a ridge waveguide geometry will be assumed in the lateral direction. All these designs could be combined with microdiscs, toroidal waveguides or any other high Q lateral waveguide designs to improve the feedback mechanisms and increase the gain in the QCL.

The original GaAs THz QCL [6] used single surface plasmon waveguides. Provided a high electrical conducting layer (e.g. a metal) can be deposited onto a dielectric (e.g. a semiconductor) and the imaginary part of the refractive index of the conducting layer is greater than the real part and also greater than that of the dielectric, electron waves called surface plasmons will be formed at the interface between the two materials and propagate along the interface (see Fig. 4.6). In the Drude theory, this is stating that the scattering time and frequency of the electromagnetic radiation is below the plasma frequency. The waveguide losses for such a waveguide are

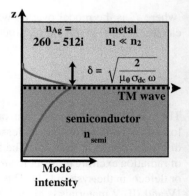

Fig. 4.6 An example of a single plasmon waveguide showing a silver metal ridge on the surface of a semiconductor. The optical modal profile is shown schematically to indicate how the mode decays into the semiconductor (not to scale)

predominantly related to the decay of the optical mode into the metal, a distance given by the skin depth of the metal. By using the Drude theory with a metal that has complex refractive index $= n_1 + i n_2$ (where $i = \sqrt{-1}$) on top of a semiconductor with refractive index n_{semi}, it can be shown that the waveguide losses for a single plasmon waveguide are approximately given by

$$\alpha_w \approx \frac{4\pi n_1 n_{semi}^3}{n_2^3 \lambda} \qquad (4.12)$$

The second trick that was used in the original GaAs THz QCL [6] to increase the modal overlap was the inclusion of a heavily doped n-GaAs layer to act both as the bottom Ohmic contact layer but also as a second plasmon reflector to increase the modal overlap with the gain region. The losses for such a layer are higher than using a metal due to the lower electrical conductivity and therefore higher skin depth than a metal.

The actual waveguide losses for each surface plasmon waveguide design require to be calculated for each structure explicitly using e.g. the Drude theory and solving the electromagnetic equations for the mode. An example for a 2.9 THz bound-to-continuum cascade design is shown in Fig. 4.7 [23]. The losses are shown as a function of different designs for the active region tuned to different frequencies for a standard

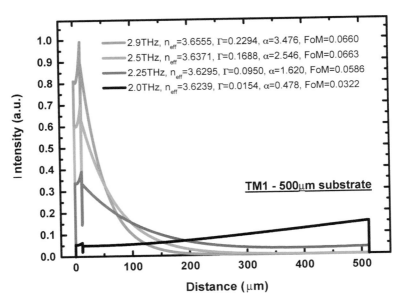

Fig. 4.7 The 1D surface plasmon mode for 2.9 (*red*), 2.7 (*green*), 2.5 (*blue*) and 2.0 (*black*) THz in an optimised 2.9 THz waveguide. The mode intensity is plotted as a function of the distance from the top of the waveguide through the bottom 600 nm n-GaAs contact and into the 500 μm thick SI GaAs substrate

waveguide of 11 μm thick active region optimised for 2.9 THz. Waveguide losses of $5\,\text{cm}^{-1}$ are computed with an overlap factor, $\Gamma = 27\,\%$. As the wavelength of the active region is increased (i.e. going to lower frequency), the skin depth of the metal plasmon waveguide decreases, reducing the intensity of the plasmon oscillations and therefore the intensity of the surface plasmon mode. The optical mode is less confined to the metal surface plasmon waveguide and the mode leaks further into the GaAs substrate thereby reducing the modal overlap but also reducing the waveguide losses as less of the mode is overlapping with free carriers. The figure of merit for waveguides is given by $\text{FOM} = \Gamma/\alpha_w$ and whilst the waveguide losses decrease, the overall figure of merit also decreases since the modal overlap decreases more quickly than the waveguide losses. This can be observed in the modal plots in Fig. 4.8 where

Fig. 4.8 The 2D TM mode profiles for the 1st, 2nd and 3rd order modes at 2.0 THz modelled using FIMWAVE 4.4 software from Photon Design [23]. The mode profile is for a 12 μm ridge waveguide active region on top of a 500 μm SI GaAs substrate. High intensity is indicated as *white* and low intensity as *black*. **a** 500 μm TM1. **b** 500 μm TM2 and **c** 500 μm TM3

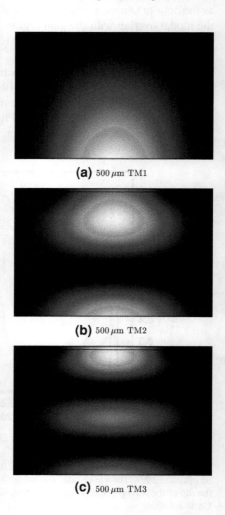

(a) 500 μm TM1

(b) 500 μm TM2

(c) 500 μm TM3

the higher order modes in the substrate become more favourable for lasing. Thinning the substrate is one way to increase the modal overlap and overall figure of merit.

The second type of waveguide is the double metal waveguide. To overcome the problems of the lower electrical conductivity of doped semiconductors compared to metals, the substrate can be etched away and a metal is deposited onto the bottom heavily doped contact layer. The waveguide is therefore a double surface plasmon with metal-semiconductor single plasmon reflectors at top and bottom to confine the optical mode vertically. This design has two major advantages. First the modal overlap is very high, typically 95 % or higher as very little of the optical mode decays into the metal if a high conductivity metal such as silver or gold is used. Secondly, the whole QCL can be directly bonded onto a copper heat-sink providing excellent heat-sinking for the device. This is especially important for high temperature operation of QCLs [16, 18, 24]. There are two disadvantages of the double metal waveguide. The waveguide losses are higher than a single plasmon waveguide as the optical mode is squeezed between the two metal layers, but this is not significant as the higher modal overlap compensates for the higher loss. The second is that the output beam is strongly diffracted as the aperture created at the facet by the two layers of metal is sub-wavelength [25].

There are other types of waveguides that have been developed in addition to those described above. The first is a photonic bandgap design which has been developed to diffract the light out of the surface of the semiconductor to produce a vertical emitter. As the TM polarization only allows edge emission, this design has the aim of allow arrays of surface emitting lasers but at the cost of reduced output power compared to the waveguides described above [26]. Microdisk cavities have also been demonstrated [27]. Such microcavities allow high Q values thereby producing much lower waveguide losses than surface plasmon ridge waveguides. The major disadvantage of such microcavities is that the output is scattered from defects, and so the output powers are again very low since the devices emit in many different directions. There are some techniques to pull the optical modes out of the microcavity in specific directions but defects still produce outputs in unwanted directions.

4.7 Exemplar Experimental Laser Results

There are now many examples of QCLs operating between around 5 THz and 800 GHz in the literature [7, 8]. As an example, the experimental results from a 2.0 THz design will be reviewed which was designed, grown and characterised in Cambridge [23]. This design uses GaAs quantum wells and $Al_{0.15}Ga_{0.85}As$ barriers and the majority of QCLs demonstrated at THz frequencies use these materials.

The band structure for the 2.9 THz (103 μm wavelength) QCL under an applied electric field of 1.5 kV/cm is shown in Fig. 4.9. This design is a bound-to-continuum QCL, where a miniband of overlapping subbands from different quantum wells is used to inject carriers into the upper laser level which is marked 2. The design consists of 8 quantum wells and 8 barriers. A thicker injection barrier of 3.8 nm (marked as i)

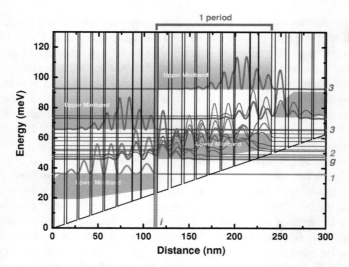

Fig. 4.9 The band structure of a 2.9 THz QCL design showing one period with the squared wavefunctions with the *green* shaded areas representing the mini bands [23]. The upper laser level is marked 2 and the lower laser level is marked 1. Also *highlighted* are the lower and upper miniband ground state g and 3. Injection into state 2 occurs via resonant tunnelling from level g through a 3.8 nm $Al_{0.15}Ga_{0.85}$ As injection barrier (i). Starting from the barrier, from left to right, the layer sequence of GaAs quantum wells ($Al_{0.15}Ga_{0.85}As$ barriers) in nm is: **3.8** / 11.6 / **3.5** / 11.3 / **2.7** / 11.4 / **2.0** / 12 / **2.0** / 12.2 / **1.8** / 12.8 / **1.5** / 15.8 / **0.6** / 9.0 / **0.6** / 14. The underlines layers are doped *n*-type to 1.6×10^{16} cm^{-3}

separates the injector from the active region where the intersubband radiative transition takes place. The highest state in the lower miniband acts as the lower laser state and electrons are then scattered down the miniband states by fast electron–electron or other fast scattering mechanisms to produce population inversion. 110 repeats of each period were sandwiched between a 70 nm thick 5×10^{18} cm^{-3} n-GaAs top contact and a 600 nm thick heavily doped bottom contact layer. Single surface plasmon waveguides were fabricated with length of 3 mm and width of 250 μm with Ohmic contacts being formed to the top and bottom using PdGe/TiAu and AuGeNi/Ti/Au, respectively before the samples were placed in a continuous flow cryostat and characterised using a Fourier Transform InfraRed (FTIR) spectrometer.

Figure 4.10 shows the current–voltage characteristics of the device as well as the electroluminescence versus current density. This QCL has a threshold current density of 115 A/cm^2 where lasing occurs. The peak power for pulsed operation is 25 mW at 4 K and corresponds to the voltage where the miniband aligns with the upper laser state. Figure 4.11 shows the spectra from the laser at different currents at 4 K. As this is a wide plasmon waveguide, more than 1 optical TM mode can be supported. The spectra show multimode emission and as the current is varied, the emission "hops" between a number of the modes in the cavity.

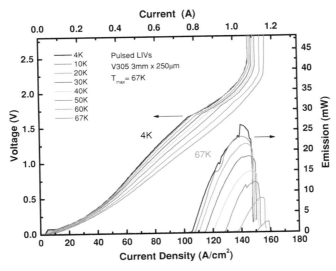

Fig. 4.10 The left axis shows the current voltage as a function of temperature for a QCL (3 mm long and 250 μm wide ridge waveguide) operating in pulsed mode with 100 μs long pulses at a repetition rate of 10 kHz. An additional 3.9 Hz 50 % duty cycle slow modulation was superimposed to match the detector response time [23]. The right hand axis shows the electroluminescence power as a function of current density

THz QCLs have also been demonstrated in a range of materials in addition to the standard GaAs/AlGaAs heterostructures. These now include InGaAs/InAlAs [28] and InGaAs/GaAsSb [29] both grown on the more expensive InP substrates. Whilst both have material parameters that suggest higher performance should be possible from these more expensive materials, the reality is that the higher band offsets result in more sensitivity to growth uniformity and reproducibility [28].

4.8 Future Requirements from THz Quantum Cascade Lasers

The experimental results from QCLs now cover an impressive frequency range, from around 800 GHz up to 5 THz with pulsed output powers up to 248 mW [17]. The majority of the lasers are produced using GaAs quantum wells and $Al_{0.15}Ga_{0.85}As$ barriers although lasers have also been demonstrated using the InGaAs system. The major problem with such THz QCLs at present is that the operating temperature is still too low for most of the applications. Figure 4.12 shows a plot of maximum operating temperature versus wavelength (frequency) for many of the reported QCLs in the literature. Also plotted is the temperature that is calculated if $k_B T$ is equated to the laser emission energy ($= \frac{hc}{\lambda}$). It is clear that the maximum operating temperature is quite close to the temperature set by the intersubband transition energy. This suggests

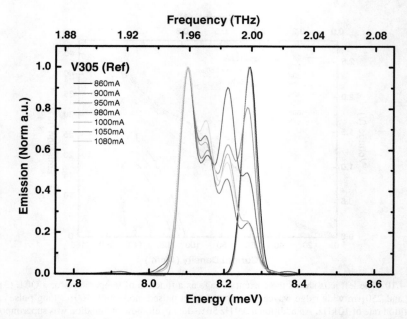

Fig. 4.11 The laser emission spectra as measured by FTIR for the 2 THz QCL for a 3 mm long and 250 μm wide ridge waveguide at 4 K in continuous wave mode. The resolution of the FTIR is 7.5 GHz and the emission lines are observed to hop between different modes in the multimode ridge waveguide as the current through the device is varied

that in the GaAs system that the thermal processes including phonon scatttering are one mechanism that are presently limiting the maximum operating temperature. Clearly, the small energy gap for the lasing transition as well as the small energy differences in other parts of the QCL design (injector and depopulation regions) make high temperature operation difficult. The highest temperature of operation of nearly 225 K was achieved in high magnetic fields [30] where the magnetic splitting of the upper and lower energy levels made many of the parasitic relaxation processes forbidden. This required magnetic fields above 10 T which are only available with superconducting magnetics that only operate at cryogenic temperatures so whilst the operating temperature is impressive, cryogenics are still required for practical applications.

The maximum operating temperature for THz QCLs without magnetic fields is 186 K [18] which is still some way from the 240 K or so required for Peltier cooler operation. The two processes that need to be addressed are the thermal phonon scattering and thermal backfilling which are shown in Fig. 4.13. Thermal phonon scattering in III-V materials is a polar process due to the electrical dipolar between the group III and group V materials. This produces a resonant process where electrons in the upper lasing subband can relax non-radiatively to the lower lasing subband in a parasitic process that does not emit light whilst emitting a LO phonon [9]. This

Fig. 4.12 A plot of experimental results from the literature of pulsed and CW QCLs showing the operating wavelength versus the maximum temperature of operation. Also plotted is a line representing the temperature expected for the energy of the laser transition

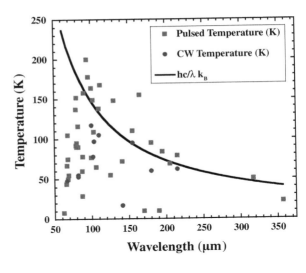

Fig. 4.13 A schematic diagram showing in green thermal backfilling from the injector of the next period (*left hand side*) and thermally activated polar phonon scattering (*right hand side*)

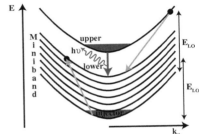

results in an exponential decrease in the upper intersubband lifetime given by

$$\tau_{\text{upper,lower}} \propto \exp\left[\frac{-(E_{\text{LO}} - h\upsilon)}{k_B T_e}\right] \tag{4.13}$$

where E_{LO} is the energy of the LO optical phonon (36 meV in GaAs) and T_e is the electron temperature in the system which is frequently much higher than the substrate temperature of the device. Monte Carlo simulations have suggested that the T_e is typically 50–100 K above the substrate temperature for phonon depopulation THz QCLs. The second mechanism is thermal backfilling where electrons are thermally excited from the next injector to the lower laser state by thermal excitation (i.e. according to the Boltzmann distribution) or through the absorption of a LO-phonon. This effect is more pronounced for bound-to-continuum designs over phonon depopulation designs and is the reason that all the high temperature operating THz QCLs have been phonon depopulation designs. Both these effects result in a reduction in the injection efficiency into the upper laser level.

Group IV materials have the advantage of non-polar optical phonons since there are no electrical dipoles between the atoms in the material [19, 31, 32]. This significantly reduces the temperature dependence of the intersubband relaxation process and indeed, lifetimes of 20 ps have been measured at 200 K where the lifetimes in GaAs quantum wells are femtoseconds [31, 32]. These non-polar LO phonon scattering processes should reduce the thermal backfilling and the parasitic thermal phonon scattering. Whilst this suggests higher operating temperatures, the lack of a polar phonon scattering mechanism actually makes the engineering of population inversion and hence gain more difficult to achieve. No lasing or gain has yet been demonstrated in the Si/SiGe material system but the most promising systems in the group IV materials appear to be with Ge quantum wells where effective masses close to GaAs values potentially allowing gain comparable to the demonstrated GaAs THz QCLs [32].

By using distributed feedback gratings on cavities, single mode THz QCLs have been demonstrated with very small linewidths. The other limitation of many THz QCLs is that the tuning of the frequency is limited to rather narrow bandwidths. The tuning of a THz QCL is related to the temperature dependence of the gain curve which is dependent on the Fermi function. This significantly limits the tuning of the devices if a design with a single frequency active region is used. The limited tuning bandwidth will always be a disadvantage of the QCL design which can only be improved through compromises with output power and/or temperature. A number of approaches are being pursued to improve the tuning including multiple active regions with different quantum well widths to increase the bandwidth along with external cavities to allow wide bandwidth tuning.

4.9 Summary

Impressive performance has been demonstrated from THz QCLs since their first demonstration in 2001 and they still represent one of the most powerful and compact coherent source of THz radiation. THz QCLs now operate with pulsed output powers up to 248 mW and cover the frequency range from around 800 GHz (375 μm wavelength) up to nearly 5 THz (60 μm wavelength). The major issue at present is increasing the operating temperature to levels where Peltier coolers could make the devices practical for the many applications that require high power THz sources. The present THz QCLs are ideal for spectroscopic applications with narrow linewidths but suffer from limited tuning range, although this has now been increased to over 1 THz of tuning. It is clear that further developments will be made to improve THz QCLs, but their present performance is already suitable for a number of the present applications, especially in astronomy telescopes where they are being used as LO sources for heterodyne detection.

References

1. P.H. Siegel, IEEE Trans. Microwave Theory Tech. **50**(3), 910 (2002)
2. M. Tonouchi, Nat. Photonics **1**(2), 97 (2007)
3. R. Kazarinov, R. Suris, Soviet Phys. Semicond. **5**(4), 707 (1971)
4. R. Kazarinov, R. Suris, Soviet Phys. Semicond. **6**(1), 120 (1972)
5. J. Faist, F. Capasso, D.L. Sivco, C. Sirtori, A.L. Hutchinson, A.Y. Cho, Science **264**(5158), 553 (1994)
6. R. Kohler, A. Tredicucci, F. Beltram, H.E. Beere, E.H. Linfield, A.G. Davies, D.A. Ritchie, R.C. Iotti, F. Rossi, Nature **417**(6885), 156 (2002)
7. B.S. Williams, Nat. Photonics **1**(9), 517 (2007)
8. G. Scalari, C. Walther, M. Fischer, R. Terazzi, H. Beere, D. Ritchie, J. Faist, Laser Photonics Rev. **3**(1–2), 45 (2009)
9. P. Harrison, *Quantum Wells, Wires and Dots: Theoretical and Computational Physics of Semiconductor Nanostructures* (Wiley Interscience, New York, 2005)
10. C. Sirtori, F. Capasso, J. Faist, A.L. Hutchinson, D.L. Sivco, A.Y. Cho, IEEE J. Quantum Electron. **34**(9), 1722 (1998)
11. H. Callebaut, Q. Hu, J. Appl. Phys. **98**(10), 104505 (2005)
12. J. Faist, F. Capasso, C. Sirtori, D.L. Sivco, J.N. Baillargeon, A.L. Hutchinson, S.N.G. Chu, A.Y. Cho, Appl. Phys. Lett. **68**(26), 3680 (1996)
13. J. Faist, M. Beck, T. Aellen, E. Gini, Appl. Phys. Lett. **78**(2), 147 (2001)
14. B.S. Williams, S. Kumar, H. Callebaut, Q. Hu, J.L. Reno, Appl. Phys. Lett. **83**(11), 2124 (2003)
15. G. Scalari, M.I. Amanti, M. Fischer, R. Terazzi, C. Walther, M. Beck, J. Faist, Appl. Phys. Lett. **94**(4), 041114 (2009)
16. B.S. Williams, S. Kumar, Q. Hu, J.L. Reno, Opt. Express **13**(9), 3331 (2005)
17. B.S. Williams, S. Kumar, Q. Hu, J.L. Reno, Electron. Lett. **42**(2), 89 (2006)
18. S. Kumar, Q. Hu, J.L. Reno, Appl. Phys. Lett. **94**(13), 131105 (2009)
19. R.W. Kelsall, Z. Ikonic, P. Murzyn, C.R. Pidgeon, P.J. Phillips, D. Carder, P. Harrison, S.A. Lynch, P. Townsend, D.J. Paul, S.L. Liew, D.J. Norris, A.G. Cullis, Phys. Rev. B **71**(11), 115326 (2005)
20. S.A. Lynch, D.J. Paul, P. Townsend, G. Matmon, Z. Suet, R.W. Kelsall, Z. Ikonic, P. Harrison, J. Zhang, D.J. Norris, A.G. Cullis, C.R. Pidgeon, P. Murzyn, B. Murdin, M. Bain, H.S. Gamble, M. Zhao, W.X. Ni, IEEE J. Sel. Top. Quantum Electron. **12**, 1570 (2006)
21. S. VisvanathanI, Phys. Rev. **120**(2), 376 (1960)
22. D.K. Schroder, R.N. Thomas, J.C. Swartz, IEEE Trans. Electron Devices **25**(2), 254 (1978)
23. C. Worrall, Long wavelength terahertz quantum cascade lasers. Ph.D. thesis, University of Cambridge, Cavendish Laboratory (2007)
24. M.A. Belkin, J.A. Fan, S. Hormoz, F. Capasso, S.P. Khanna, M. Lachab, A.G. Davies, E.H. Linfield, Opt. Express **16**(5), 3242 (2008)
25. A.J.L. Adam, I. Kasalynas, J.N. Hovenier, T.O. Klaassen, J.R. Gao, E.E. Orlova, B.S. Williams, S. Kumar, Q. Hu, J.L. Reno, Appl. Phys. Lett. **88**(15), 151105 (2006)
26. L. Mahler, A. Tredicucci, Laser Photonics Rev. **5**(5), 647 (2011)
27. G. Fasching, A. Benz, R. Zobl, A. Andrews, T. Roch, W. Schrenk, G. Strasser, V. Tamosiunas, K. Unterrainer, Physica E Low Dimens. Syst. Nanostruct. **32**, 316 (2006)
28. M. Fischer, G. Scalari, K. Celebi, M. Amanti, C. Walther, M. Beck, J. Faist, Appl. Phys. Lett. **97**(22), 221114 (2010)
29. C. Deutsch, A. Benz, H. Detz, P. Klang, M. Nobile, A.M. Andrews, W. Schrenk, T. Kubis, P. Vogl, G. Strasser, K. Unterrainer, Appl. Phys. Lett. **97**(26), 261110 (2010)
30. A. Wade, G. Fedorov, D. Smirnov, S. Kumar, B.S. Williams, Q. Hu, J.L. Reno, Nat. Photonics **3**(1), 41 (2009)
31. M. Califano, N.Q. Vinh, P.J. Phillips, Z. Ikonic, R.W. Kelsall, P. Harrison, C.R. Pidgeon, B.N. Murdin, D.J. Paul, P. Townsend, J. Zhang, I.M. Ross, A.G. Cullis, Phys. Rev. B **75**(4), 045338 (2007)
32. D.J. Paul, Laser Photonics Rev. **4**(5), 610 (2010)

Chapter 5
Relativistic Electrons-Based THz Sources: Principles of Operation and the ENEA Experience

Andrea Doria

Abstract An introduction to the general theory of free electron sources of coherent radiation is reported. The basic mechanisms for the generation of electromagnetic waves as spontaneous emission process are described together with the more sophisticated theory of gain for the description of the stimulated process. The ENEA experience in the realisation of compact free electron sources is reported jointly with the "tricks" that allowed the Frascati group to shorten the device dimensions while maintaining important performances.

Keywords Free electron laser · Undulator · Terahertz source · Compact FEL · Synchrotron radiation, Cerenkov FEL · Smith-Purcell FEL

An introduction to the general theory of free electron sources of coherent radiation is reported. The basic mechanisms for the generation of electromagnetic waves as spontaneous emission process are described together with the more sophisticated theory of gain for the description of the stimulated process. The ENEA experience in the realisation of compact free electron sources is reported jointly with the "tricks" that allowed the Frascati group to shorten the device dimensions while maintaining important performances.

5.1 The Electron–Photon Interaction

The Free Electron Laser (FEL) is a very interesting device with many peculiar characteristics that makes it unique [1–3]. It is essentially based on the interaction between a free electron and a photon. It is well-known that a single electron cannot emit or absorb a single photon in vacuum, due to the fact that energy and momentum con-

A. Doria (✉)
Laboratorio Sorgenti, UTAPRAD, ENEA, Via Enrico Fermi 45, 00044 Frascati (Rome), Italy
e-mail: andrea.doria@enea.it

M. Perenzoni and D. J. Paul (eds.), *Physics and Applications of Terahertz Radiation*, Springer Series in Optical Sciences 173, DOI: 10.1007/978-94-007-3837-9_5, © Springer Science+Business Media Dordrecht 2014

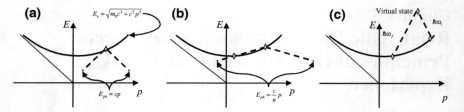

Fig. 5.1 Phase space: **a** in vacuum emission; **b** Cerenkov emission; **c** Compton scattering

servation laws cannot be violated. This fact is evident looking at the Phase-Space reported in Fig. 5.1a where the dispersion relation of a free relativistic electron is reported; as can be seen if an electron is found in its initial state for which it is asked to emit a photon, it is required that the final energy-momentum state of the photon must lie on the cone represented by its dispersion relation. Since the same energy and momentum must be carried off from the electron final state, it is required that the two dispersion relation curves must intercept to let this process occur. In vacuum this will never happen. But if one moves the problem from vacuum to a medium having a refractive index n, the photon dispersion relation changes because now an interception occurs (see Fig. 5.1b), that corresponds to the final state of the electron, allowing an emission process called the Cerenkov effect. This is not the only emission mechanism permitted, the most exploited one is related to the Compton scattering. This process can be fully understood in a quantum picture, within which a simultaneous process of emission and absorption of photons can be explained by means of a virtual state (see Fig. 5.1c); an intense magnetic field, in a relativistic regime, can be regarded as a virtual photon and, from the energy-moment conservation laws, the absorbed and emitted photon are simply connected by the electron kinetic energy.

Let us limit to a classic picture and the discussion about the main mechanism responsible for the radiation emission in Free Electron Devices. The basic mechanism is the synchrotron radiation emission to which an electron undergoes when passing through a bending magnet, and thus performing a bent trajectory; due to the central acceleration, the electron emits radiation in a very short pulse related to the electron transit time along an arc section (see Fig. 5.2). The correspondent emitted spectrum, calculated by means of the Lienard-Wiechert potentials, is very broad and has a long tail toward the high-frequency region [4]. The characteristic frequency of such emission is: $\omega_c = (3c)/(2\rho) \cdot (E/m_o c^2)^3$ and result to be proportional to the cube of the electron energy E and inversely proportional to the orbit radius ρ (being c the light velocity and m_0 the electron rest mass; see Fig. 5.2a).

In a magnetic undulator [5], due to the alternate arrangement of the magnetic poles, the electron undergoes oscillations around the symmetry axis. If the angle $\theta = K/\gamma$ of the sinusoidal trajectory is smaller than the light cone angle $(1/\gamma)$, than the observer can see a pulse longer in duration with respect to the previous case. This is realised when the so-called undulator parameter K is smaller than one $(K = \left(e \langle B \rangle^{0.5} \lambda_u\right) / \left(2\pi m_o c^2\right))$. This parameter is directly related to the mag-

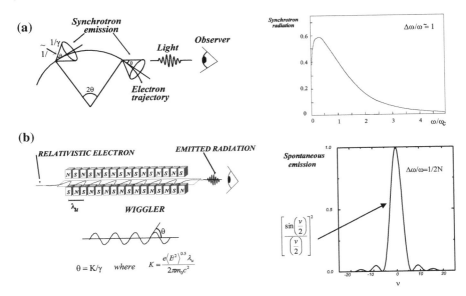

Fig. 5.2 Synchrotron radiation emission

netic field strength B and to the undulator period λ_u. The synchrotron emission in a magnetic undulator results in a narrower emission spectrum that has a relative bandwidth that is inversely proportional to the number of undulator periods N (see Fig. 5.2). The shape of the emitted radiation has a "sinc" behaviour and this is simply related to the finite length of the magnetic devices. In contrast to the bending magnet synchrotron radiation, for which the number of photons N_f emitted is directly related to the number of electrons N_e involved in the curved trajectory ($N_f \sim N_e$), the presence of a complex magnetic structure changes the properties of the emitted radiation. In a wiggling device (an undulator with a parameter $K > 1$), the number of photons is amplified by the number of oscillations (the number of periods N) but no positive interference occurs among the photons generated and thus $N_f \sim N_e N$. If a magnetic device with an undulator parameter $K < 1$ is used instead of a wiggler, the interference effects arise among emissions at any bend and the radiation is produced by a large number of essentially collinear source points with the result $N_f \sim N^2 N_e$.

5.2 The Free Electron Laser Mechanism

The main elements constituting a FEL are essentially three: The first element is the electron accelerator. The accelerated particles are transported through an electron transport channel, by means of optics elements like steering coils and quadrupoles.

Electrons are transported inside the interaction region, that is the second element of the device, (i.e. a magnetic undulator, a surface waveguide, a metallic grating etc.), where their kinetic energy is transformed into radiation field; this is what is called, in the laser-physics glossary, spontaneous emission. The third element is the optical cavity where the radiation field is stored. The emission mechanism can be stimulated by the presence of the stored radiation. The result is an increase of the output power due to the simultaneous interaction among the electrons, the electric field of the stored radiation and the physical element peculiar for the interaction region (magnetic field in an undulator, longitudinal surface wave in a dielectric waveguide, the image charge field for the metallic grating etc.).

5.2.1 Spontaneous Emission

The central emitted frequency can be calculated by the synchronism condition that, in the case of an undulator FEL, operating in vacuum, is realised when an electron with a normalised energy γ, after one undulator period λ_u, having a magnetic peak intensity B, oscillates with a phase difference of 2π with respect to the electromagnetic wave, i.e. $(\omega t - kz) = 2\pi$ at $z = \lambda_u$ and $t = \lambda_u/(c\beta_z)$ being β_z the longitudinal normalised velocity; the result is a beam line equation $\omega/(c\beta_z) - k - k_u = 0$ that, in the free space, where $k = \omega/c$, can be arranged as: $\omega/c = \beta_z(1 + \beta_z)\gamma_z^2 k_u$. Introducing the longitudinal normalised energy as: $\gamma_z^2 = \gamma^2/(1 + K^2)$ we obtain the fundamental equation for a FEL [1–3]:

$$\lambda_0 = \frac{\lambda_u}{2\gamma^2}(1 + K^2) \quad \text{where} \quad K = \frac{e\left\langle B^2\right\rangle^{0.5}\lambda_u}{2\pi m_0 c^2} \tag{5.1}$$

This equation indicates that the emitted wavelength is directly proportional to the undulator period λ_u and inversely proportional to the square of the electron energy γ; it is also weakly proportional to the magnetic field B by means of the undulator parameter K. All these parameters can be established during the design project and some of them can be continuously varied, the result is a continuous tuning of the output frequency. This is the most appealing peculiarity of an FEL if compared with conventional lasers where the emitted wavelength is strictly related to quantum level energy differences.

As already discussed about the spontaneous emission spectral line-shape it can be only added that, being the emission process not related to quantum levels, the natural line-width $\Delta\omega$, in the homogeneous broadening regime, is only related to the time finiteness of the process due to the electron transit time δt in the interaction region of length L (magnetic undulator, dielectric-loaded waveguide, metal grating etc.) and to the velocity difference between electrons v_z and photons c.

$$\delta t = \frac{L}{v_z} - \frac{L}{c} \approx \frac{L}{2c} \frac{1}{\gamma^2} \left(1 + K^2\right) \quad \Rightarrow \quad \Delta\omega \approx \frac{\pi}{\delta t} \frac{2\pi c}{L} \gamma^2 \frac{1}{\left(1 + K^2\right)} \qquad (5.2)$$

Combining Eq. (5.2) with Eq. (5.1) one obtains the relative bandwidth for a magnetic undulator-based FEL: $\Delta\omega/\omega = 1/(2N)$, that results to be inversely proportional to the number of periods N. The radiation pulse with time duration and bandwidth as reported in Eq. (5.2) implies a line-shape given by:

$$f(\nu) \propto \left[\frac{\sin(\nu/2)}{\nu/2}\right]^2 \quad \text{where} \quad \nu = 2\pi N \frac{\omega_0 - \omega}{\omega_0} \qquad (5.3)$$

where ω_0 is the central emitted frequency obtained from Eq. (5.1).

If one wants to operate a free electron laser at long wavelengths it is needed to use guiding structures capable to confine the radiation that otherwise suffers high diffraction; the FEL resonance condition can thus be studied in the presence of a rectangular waveguide [6]. In this case, the guiding structure has a dispersion relation different with respect to the free-space one ($\omega/c = \sqrt{\Gamma_\lambda^2 + k^2}$ where $\Gamma_{\lambda(m,n)} = \sqrt{(m\pi/a)^2 + (n\pi/b)^2}$ and a and b indicate the transverse waveguide dimensions); the dispersion relation of the fundamental TE_{01} mode, in a rectangular waveguide, is plotted in Fig. 5.3 together with some beam-lines and the light-line of unity slope (dotted line). Generally speaking, the dispersion relation of a generic TE_{0n} mode intercepts the ω/c axis at the cut-off frequency $(\omega/c)_{co} = \Gamma_{0n}$, and, due to the transverse oscillations of the electrons induced by the magnetic field associated force of the undulator, the beam line (i.e. expressed by means of the detuning parameter $\theta \equiv \delta k \cdot L = (\omega/(c\beta_z) - k - k_u)L$) also intercepts the (ω/c) axis at a frequency corresponding to a value $\beta_z k_u$ where k_u is the undulator momentum $(2\pi/\lambda_u)$ and has a line coefficient equal to β_z. In a general situation, there are now two distinct solutions corresponding to the interceptions between the dispersion relation and the beam line, giving rise to two resonant frequencies:

$$\frac{\omega_\pm}{c} = \beta_z \gamma_z^2 k_u \{1 \pm \beta_z \Delta\} \quad \text{with} \quad \Delta = \left[1 - \left(\frac{\Gamma_{0n}}{\beta_z \gamma_z k_u}\right)^2\right]^{1/2}, 0 \leq \Delta \leq 1 \qquad (5.4)$$

The parameter Δ characterises the shift of the resonant frequencies with respect to the free-space resonance corresponding to Eq. (5.1). Analysing Eq. (5.4) we may observe how, by a proper choice of the undulator parameters, waveguide dimensions and electron beam energy, the two solutions can move with respect to each other. We will come back to this point in Sect. 5.4.2).

Fig. 5.3 A graphical representation of the synchronism condition for a waveguide undulator FEL

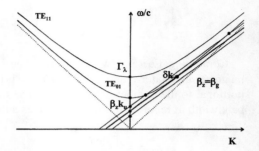

5.2.2 Gain

The theory of the gain mechanism in free electron lasers is a quite elegant one and it is worth to be described in its essential elements. As anticipated in the introduction of Sect. 5.2, stimulated emission occurs when electrons exchange energy in favour of the radiation field with a proper phase relation. So the problem of understanding the gain mechanism reduces to the evaluation of this amount of energy exchanged between the electrons and the radiation field. This quantity can be calculated as follows: Let define the gain as the ratio between the electron–radiation energy exchange with respect to the radiation energy stored in the cavity [7].

$$G \equiv \frac{\Delta W_L}{W^0{}_L} = -m_0 c^2 \frac{\langle \Delta \gamma \rangle}{W^0{}_L} \tag{5.5}$$

Where $W^0{}_L$ is the electromagnetic energy stored in the optical cavity. On the other hand, the electron–radiation energy exchange ΔW_L is given by the following expression that integrates, over the volume and over the transit time duration in the undulator, the scalar product of the current density with the electric field associated with the radiation.

$$\Delta W_L = -\int_0^{T_L} \int_V \mathbf{J}(\mathbf{x}, t) \cdot \mathbf{E}(\mathbf{x}, t) d\mathbf{x} \tag{5.6}$$

Where $\mathbf{J}(\mathbf{x}, t)$ represents the electron current density and $\mathbf{E}(\mathbf{x}, t)$ is the electric field associated to the electromagnetic wave. An explicit evaluation of the expression leads to the following result: $\int_V \mathbf{J} \cdot \mathbf{E} d\mathbf{x} = I (L/(c\beta_z)) E_L v_T$ where I is the electron current, the electric field $E_L = E_0 \cos(\omega t - kz + \phi_L)$ shows a sinusoidal behaviour and the transverse electron velocity $v_T = -\left(cK\sqrt{2}/\gamma\right) \sin(2\pi z/\lambda_u)$ contains the undulator parameters, because it is the magnetic field of the structure that induces it. Please note that the sinusoidal behaviour of the electron velocity with a periodicity is equal to the undulator period, and also that the coupling is only a transverse coupling.

The normalised rate of energy exchange expressed as the time variation of the relativistic factor γ is thus expressed as follows:

$$\dot{\gamma} = \frac{eE_0K}{m_0c\gamma\sqrt{2}}\sin(\psi) \quad \text{where} \quad \psi = \omega t - \left(k + \frac{2\pi}{\lambda_u}\right)z + \phi_L \quad (5.7)$$

where the normalisation is given by the electron rest energy m_0c^2. The phase term takes into account the phase relation between the electron and the electromagnetic wave.

In order to obtain the basic equation that regulates the dynamics of an FEL, we need to evaluate the trajectory of the single electron in the undulator. The electron velocity components add in a quadratic way (any vertical contribution has been neglected) the transverse velocity refers to an average value and the total velocity is expressed in terms of the relativistic factor γ. With the previous expressed value of the transverse velocity one obtains for the longitudinal velocity the following relation:

$$\dot{z}^2 + \left\langle v_T^2 \right\rangle = c^2\left(1 - \frac{1}{\gamma^2}\right) \Rightarrow \dot{z} = c\sqrt{1 - \frac{\left(1 + K^2\right)}{\gamma^2}} \quad (5.8)$$

that with simple algebra leads to the following equation of motion:

$$\ddot{z} = c\frac{\dot{\gamma}}{\gamma^3}\frac{\left(1 + K^2\right)}{\beta_z} \quad \text{where} \quad \beta_z = \sqrt{1 - \frac{1}{\gamma_z^2}} \quad \text{and} \quad \gamma_z = \frac{\gamma}{\sqrt{1 + K^2}} \quad (5.9)$$

By the use of Eq. (5.7) substituted in Eq. (5.9), the expression that relates the longitudinal acceleration with the phase term ψ regulating the electron–radiation interaction as stated in Eq. (5.7), is obtained:

$$\ddot{z} = \frac{eE_0K\left(1 + K^2\right)}{m_0\gamma^4\beta_z\sqrt{2}}\sin(\psi) \quad (5.10)$$

But from the definition of ψ, again expressed in Eq. (5.7), a differential equation for the phase variable $\ddot{\psi} = -\left(k + 2\pi/\lambda_u\right)\ddot{z}$ is achieved.

$$\ddot{\psi} = -\Omega^2\sin(\psi) \quad \text{where} \quad \Omega^2 = \left(k + \frac{2\pi}{\lambda_u}\right)\frac{eE_0K\left(1 + K^2\right)}{m_0\gamma^4\beta_z\sqrt{2}} \quad (5.11)$$

The form of the equation is that of a classical pendulum equation where the Rabi frequency (that regulates the Bloch oscillations in two quantum level systems) is proportional to the intracavity radiation amplitude, by means of the associated electric field E_0, and by the magnetic undulator parameters as the magnetic field amplitude and period, by means of the parameter K.

This is quite an impressive and elegant result, but the solution of such an equation goes beyond the aim of this introduction to FEL physics: Anyway a reading of the result can be given. Going a step back and combining Eq. (5.10) with the $\ddot{\psi}$ expression, derived from Eq. (5.7), we get:

$$\ddot{\psi} = -\left(k + \frac{2\pi}{\lambda_u}\right) c \frac{\dot{\gamma}}{\gamma^3} \frac{\left(1 + K^2\right)}{\beta_z} \tag{5.12}$$

From which, by a proper integration, we get:

$$\Delta \gamma = -\frac{\gamma^3 \beta_z}{c \left(k + \frac{2\pi}{\lambda_u}\right) \left(1 + K^2\right)} \Delta \dot{\psi} \Rightarrow G = \frac{\gamma^3 \beta_z \left(m_0 c^2\right)}{c \left(k + \frac{2\pi}{\lambda_u}\right) \left(1 + K^2\right) W^0{}_L} \langle \Delta \dot{\psi} \rangle \tag{5.13}$$

The gain function derived in Eq. (5.13) has been obtained by the use of the definition reported in Eq. (5.5). Moreover, an average $\langle \Delta \dot{\psi} \rangle$ has to be performed over the electron distribution. This procedure is important because the net gain from an electron bunch is needed, and thus the average contributions of all the electrons each carrying its phase relation.

In order to calculate this phase averaged quantity, that is a fundamental term of the FEL dynamics, we need to solve the pendulum equation (5.11) by means of small perturbation techniques under the hypothesis of a small signal and small gain regime that will ensure a small value for the Rabi frequency (5.11) i.e. $\Omega << c/L$. The result is a gain function showing a behaviour $g(\theta)$ as a function of frequency, that can be considered "universal" for all free electron-based radiation sources [6, 7]:

$$\langle \Delta \dot{\psi} \rangle = \left(\frac{\Omega L}{c}\right)^4 \frac{c}{4L\beta_z^3} \underbrace{\frac{d}{d\theta} \left[\frac{\sin(\theta/2)}{\theta/2}\right]^2}_{g(\theta)} \Rightarrow G = \pi \frac{\left(1 + K^2\right) K^2}{\beta_z^5 \gamma^5} \frac{\lambda_u^3 N^3}{\Sigma_L} \frac{I}{I_0} \left(k + \frac{2\pi}{\lambda_u}\right) g(\theta) \tag{5.14}$$

where Σ_L is the radiation mode cross section and I_0 is the Alfven current [6, 7].

It is important to underline the behaviour of the gain respect to some important design parameters; first of all, the electron beam current I shows a linear trend in the gain, while the undulator length contributes with the third power by means of the period number N. This means that to increase the gain we need a high current electron beam and, much more easily due to the cubic power, a long undulator device. Another important parameter that can be set during the design of an FEL is the electron energy, expressed by the relativistic factor γ, for which we have an inverse proportionality to the fifth power, indicating that high-energy electron beams, needed to operate FELs at short wavelengths (see Eq. (5.1)), severely reduce the gain and the possibility to design such a laser source from the visible spectral region down to ultraviolet and beyond, unless very long undulators are used.

We can conclude this brief introduction to the gain mechanism in FEL devices discussing about the previous mentioned universal spectral line-shape $g(\theta)$ (where θ is the generalised expression, for radiation generation in guiding structures, of the detuning parameter reported in Eq. (5.3)). The first thing to be noticed is that the gain frequency dependence is proportional to the derivative of the spontaneous emission line shape that is a "sinc" function as plotted in Fig. 5.4; this behaviour is general for all the free electron-based lasers and is known as Madey's theorem [8].

Fig. 5.4 The universal
spectral line-shape for free
electron-based lasers

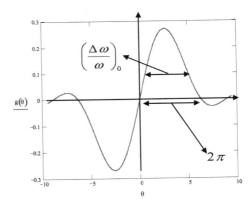

Moreover, the θ parameter is directly related to the ψ parameter, expressed by
Eq. (5.7), and can be quoted in a simple form at least for the free space operation
for which we can read the frequency shift with respect to the spontaneous emission
resonant frequency as already described in Sect. 5.2.1:

$$\theta = L\left(\frac{\omega}{c}\frac{1}{\beta_z} - k - k_u\right) = L\delta k \rightarrow 2\pi N\frac{\omega - \omega_0}{\omega_0} \tag{5.15}$$

Another interesting consideration to be proposed, reading Eq. (5.14) and looking
at Fig. 5.4. is that the gain at $\theta = 0$ i.e. at the resonant frequency, is equal to zero; this
means that there is a slight shift in frequency, between spontaneous and stimulated
emission for the FELs, due to the Madeys theorem. In the presence of a guiding
structure, we have to deal with a dispersion relation between the wavenumber k and
the frequency ω, but also in this case the phase-shift θ would be proportional to the
momentum shift δk between the beam-line and the dispersion relation of the guide,
as reported in Eq. (5.15) and visible in the phase-space drawn in Fig. 5.3. Finally,
the bandwidth of the gain curve can be easily calculated from the sinc line shape
characteristics and the result is inversely proportional to the undulator number of
periods and thus to its length [7].

$$\left(\frac{\Delta\omega}{\omega}\right)_0 = \frac{1}{2N} \Rightarrow \Delta\omega_0 = \frac{\pi c}{N\lambda_0} \tag{5.16}$$

5.3 COMPACT FELs

In order to compose a design for an FEL it must be said that there are practical
limitations on the production of magnetic undulators with small periods and high
magnetic fields, suitable for high gain operation at short wavelengths (see Eq. (5.14)).

Table 5.1 Electron beam sources [9]

Accelerator	Energy	I-Peak	Pulse duration	Spectral region
Pulse diode	<2 MeV	1–10 kA	0.1–1 μs	mm-wave
Electrostatic	<10 MeV	<10 A	1–100 μs	THz, mm-wave
Microtron	<30 MeV	<10 A	15–30 ps (1–50 μs macropulse)	IR, THz, mm-wave
Induction linac	<50 MeV	10 kA	100 ns	IR, THz, mm-wave
RF linac	2 MeV–1 GeV	10 A–1 kA (compressed)	1–30 ps (1 – 50 μs macropulse)	IR, Visible, UV, XUV
Storage ring	0.2–2 GeV	100 A	0.1–1 ns	Visible, UV, XUV

This means that an FEL project requires to set first the parameters related to the electron beam and thus it is important to make a proper choice about the accelerator source, depending on the needed characteristics of the radiation to be generated. In the table below are listed the accelerators most frequently used, as electron beam source, for FELs together with some of their characteristics.

In the last column of Table 5.1, it has been reported that the spectral range of operation of an FEL utilising the correspondent accelerator in column one, considering, as interacting structure, an undulator, with a typical period length of few centimetres, and an on-axis magnetic field of some kilo-gauss. It is important to notice how different is the time structure for the different accelerators, especially because it must be remembered that this structure is almost exactly replicated to that of the radiation. As it can be deduced, the realisation of an FEL in the visible and Near InfraRed (NIR) spectral regions, requires the use of high energy electron beams that can be generated only by Radio Frequency (RF) Linacs or Storage Ring that are usually huge devices and very complex systems to operate. It must be also stressed that in these spectral regions very few applications cannot be satisfied by conventional laser sources. On the contrary, the spectral region ranging from beyond 30 μm up to several millimetres, that is conventionally indicated as the Terahertz region, still offers many possibilities to realise free electron-based sources with particular power or time structure characteristics.

All the elements are now available to discuss how to proceed for the design of a compact FEL source. From the analysis of Eq. (5.1) and Table 5.1 it is clear how small energy electron accelerators are suitable for long wavelength FEL operations, but, from the Eq. (5.14), low-energy electrons means also higher gain (gain is inversely proportional to the fifth power γ). The result is that long wavelength FELs, may take benefit by the use of reduced size accelerators. The enhanced gain allows the use of shorter interaction regions having thus the possibility to increase the efficiency that is linearly dependent by the undulator length [1–3]; moreover, interacting structures different from the undulator, as the slow wave-guided structures, can be used. At long wavelengths a category of mirrors can be used, like metal meshes or wire polarisers, that are transparent to the electrons; this reduces the resonator length and the electrons transport channel avoiding the use of insertion and extraction magnets. All these

Fig. 5.5 ENEA Compact
FEL facility

considerations demonstrate that there is a tight connection between Millimetre (MM) wave or THz FEL and the compactness of the source that can be realised.

5.4 The ENEA Frascati FEL Facility

In the ENEA Frascati laboratories, since the late 1980s, we have studied the problems related to the long wavelength FEL operation and construction; the result is the realisation of a FEL facility (see Fig. 5.5), based on a microtron powered by a 2.5 MW magnetron as the Radio-Frequency source and with a straight line electron transport channel and a vacuum chamber designed to host different interacting structures. The microtron is a very compact and versatile accelerator; it uses a single RF cavity and a magnetic field that forces the electrons to circular orbits up to the iron extraction pipe. Its final energy can be changed just by moving the RF cavity with respect to the extraction pipe thus changing the number of orbits and the correlated energy gain. The electron energy can be varied from 2.3 to 5.5 MeV. The peak current ranges from 6 to 4 Amps over a bunch of 15 ps. The macropulse duration is 4 μs (see Table 5.2 for the main parameters). The emittance and energy spread may look not to be excellent, but they are suitable for long-wavelength operation. The most appealing part of the device is the vacuum chamber in which where hosted different interacting structures like a short magnetic undulator for the compact FEL operating in the MM wave range, or a dielectric-loaded waveguide for a MM and sub-MM wave Cerenkov FEL and, finally, a metal grating for a Smith-Purcell based FEL that operated, with low emitted power in a wide spectral range [10].

Table 5.2 ENEA Compact FEL Parameters

Electron energy [MeV]	2.3–5
Peak current [A]	6-4
Average current [A]	0.25-0.2
Micropulse duration [ps]	15
Macropulse duration [μs]	4
Normal emittance [cm rad]	0.02
Energy spread	1%

5.4.1 The Cerenkov FEL Experiment

The first experiment realised in the facility was the Cerenkov-based FEL. As mentioned in Sect. 5.1, if we have a dielectric medium the radiation can be slowed down by a factor ε, corresponding the dielectric constant, giving to the electron the possibility to synchronise with the radiation field and allow the emission. During such a process electrons passing at grazing incidence to the surface of a dielectric film deposited on a metallic substrate, couple to the longitudinal electric field of a TM-like surface mode of the waveguide that has an evanescent behaviour in the transverse direction (see Fig. 5.6a). This longitudinal coupling introduces differences, in the synchronism mechanism, with respect to what exposed in Sect. 5.2.1) for the more conventional transverse coupling. The dispersion relations of this slow-wave structure is a quite complicated one due to the presence of the dielectric and can be calculated imposing to the magnetic and electric components of the radiation field, the boundary and continuity conditions [11, 12]. The plot of this relation in the phase space is reported in Fig. 5.6b, and results to be the solution of the indicated four equations system. Synchronism is realised when the electron velocity and the phase velocity of the radiation mode equals: i.e $\beta = \omega/(ck)$. Gain is positive, as for the undulator FEL, in the θ positive region; this is the region where the electron velocity exceed the phase velocity of the radiation, giving the well-known Cerenkov condition. It has to be noticed that for $\theta > 0$ the electron velocity exceed also the radiation group velocity ($v_g = d\omega/dk$). The dispersion relation and the electron beam line interception give the following result for the radiation emitted wavelength:

$$\lambda = 2\pi d\gamma \frac{\varepsilon - 1}{\varepsilon} \tag{5.17}$$

where d is the thickness of the dielectric film, $\varepsilon = n^2$ is the dielectric constant and γ is the relativistic factor of the electrons. The main difference between Eq. (5.17) and Eq. (5.1) is evident: The radiation wavelength is now directly proportional to the electron energy and not inversely as for the transverse-coupled undulator FEL. The result is that relatively short wavelength can be reached with low-energy accelerators.

For the ENEA Cerenkov FEL [13], the microtron energy has been set to 5 MeV and both a single and double slab waveguide geometries have been used. A hybrid waveguide resonator has been realised, having a length of 30cm, by means of diamond-machined copper, with a perfectly polished surface, and by a polyethylene film, of different thickness deposited over it (see Fig. 5.7a). The resonator was realised using a mesh grid reflector as the output coupler, made of 500 l/inch, and a cylindrical mirror, copper made, with 45 cm of radius. A miniaturised e-beam injection scheme composed of two pairs of permanent magnets, placed symmetrically sideways at the entrance of the resonator, allowed the e-beam to enter the resonator above the input mirror and then be displaced so as to travel parallel and close to the surface of the dielectric film. The radiation was then coupled into a horn and reflected out by a 2000 l/inch metal mesh. In agreement with the excitation of TM-like surface waves, the correct polarisation of an emitted radiation was measured by means of tungsten wire polarisers and the maximum output was observed when the e-beam centroid was placed close to the surface of the film acting on the magnets distance; this operation has been done when operating with the single-slab geometry. The double-slab device (Fig. 5.7b) does not require such a highly accurate alignment because the field exponential decay, related to the evanescent behaviour of the guided mode, is compensated by the presence of the upper metal plate.

Two different polyethylene films, with thickness 25 and 50 μm, respectively, were successfully tested during the experiment. The wavelength of the emitted radiation was 1660 μm for the 50 μm-thick film and 900 μm for the 25 μm-thick film. A quasi-optical resonator provided feedback and confinement of the radiation.

The spectra of the Cerenkov FEL emission, measured with a far-infrared Fabry–Perot interferometer, are reported in Fig. 5.8 for both films. The Fabry-Perot utilises, as mirrors, some metal meshes with spacing of 200 and 250 μm, respectively. The estimated Finesse was $F = 20$ for $\lambda = 1.66$ mm and $F = 10$ for $\lambda = 0.9$ mm. Maximum power was obtained with the 50 μm-thick film; running the accelerator at 10 Hz, an average power of 80 mW was measured by means of a disc calorimeter. Taking into account the coupling losses of the collecting system this corresponds to 50 W of emitted power over the 4 μs macropulse [13].

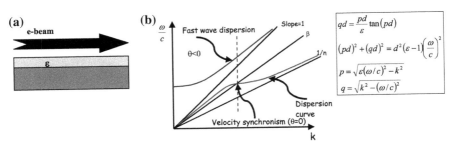

Fig. 5.6 The Cerenkov FEL Interaction scheme (**a**) and Phase-Space dispersion relation (**b**)

Fig. 5.7 The Cerenkov Waveguide Hybrid Resonator: **a** single-slab geometry; **b** double-slab geometry

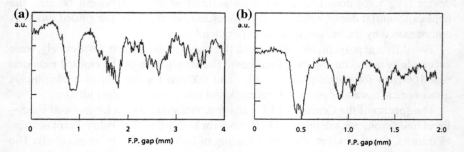

Fig. 5.8 Fabry-Perot Interferograms for the ENEA Cerenkov FEL experiment: **a** film thickness experiment d = 50 μm (4 mm scan–10 μm step) **b** film thickness experiment d = 25 mm (2 mm scan–10 μm step)

5.4.2 The Compact Undulator FEL Experiment

After the Cerenkov-based source experiences the ENEA FEL team decided to use a compact undulator for realising a different and more powerful FEL source. The long wavelength spectral region, from mm-wave to THz, offers the possibility of using the unique combination of low-energy accelerators, small-size undulators and waveguide hybrid resonators with which a compact FEL device can be realised. As mentioned at the beginning of Sect. 5.4, these devices, due to the long operating wavelength, the use of RF accelerators and the presence of guiding structures, introduce new features with respect to traditional FELs. First of all, the possibility of having a signal built up from coherent emission at the RF harmonics, as discussed in the following.

The use of a guiding structure [6], inside the interaction region, introduces some advantages with respect to the free space operation. The optical mode with a smaller and constant cross-section increases the gain because it ensures the best overlapping between the electrons and the field (see Eq. (5.14)). The gain enhancement allows the

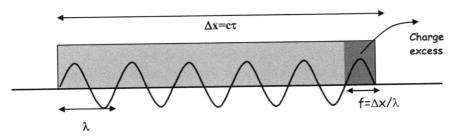

Fig. 5.9 An electron beam microbunch

use of compact size undulators. Moreover with low e-beam energies, metal meshes can be used as electron transparent resonator mirrors with an acceptable degradation of the electron beam quality; such optical cavities do not require the use of large magnets for the insertion and extraction of the electrons. All these advantages go in the direction of reducing the overall sizes of the FEL device for the FIR and MM wave operation.

Before describing the experimental results it is worthwhile to stress how the time structure of RF FEL is relevant to some results; in particular, it will be evident how the order contained in the electron beam results in a coherence of the emitted radiation. During the Cerenkov FEL experiment, we realised that a considerable amount of power was detected even when the cavity was misaligned. After many tests and discussions, we argued that an intrinsic property of the bunch was responsible for this anomalous emission with respect to the theory; this property is the "coherence". It is now well-known that, in RF accelerators, if the bunch length of an electron beam is small enough to be comparable to the radiation wavelength, we could always have an excess of phase, and thus an excess of charge, for which the emission of the different parts cannot be compensated. This fraction $f = \Delta x / \lambda$ is responsible to a coherence in the generation mechanism that goes with the square of the number of electrons in this fraction (see Fig. 5.9) [14].

The radiated power is estimated to be $P \approx N + (fN)^2$ (N is the number of electron in a bunch; typically, $N \approx 10^8$) where the first term is related to the single particle emission and the second to the coherent one. The coherent emission dominates when $f^2N >> 1$ i.e. when the fraction of the charge in excess is, at least, of the order of 10^{-4}. Comparing it with the length of the bunch Δx, the result is that coherent emission dominates in the FIR or at longer wavelengths.

Another useful tool to be exploited is the coherence among the bunches. A RF accelerator generates a train of bunches, and if the correlation among consecutive bunches is good, the radiation will be emitted at discrete frequencies which are harmonics of the RF. This experimental evidence is related to the fact that, as well-known, a RF accelerator generates a train of bunches (microbunches) that is usually as long as several microseconds (see Fig. 5.10). This time structure allows us to perform an expansion of the electron transverse current $\boldsymbol{J}_T$ into the Fourier spectral components that are the harmonics of the fundamental frequency ν_{RF}[15]:

$$\vec{J}_T = \sum_{l}^{\infty} \vec{J}_{Tl} \exp\left(-i\omega_l t\right) \quad \text{where} \quad \omega_l = 2\pi \nu_{RF} l \qquad (5.18)$$

The emission mechanism does not modify the spectral components of the e-beam, but, if the correlation among consecutive electron bunches is good, also the radiation will be emitted at discrete frequencies which are harmonics of the RF; this fact happens because every Fourier term of the current expansion in Eq. (5.18) acts as a single current source that couples with the electric field of the radiation, expressed as the transverse mode of the waveguide. The energy exchange between the guided mode λ and the harmonic l is expressed by the coefficient $A_{l,\lambda}$ (see Eq. (5.19) where Z_0 is the vacuum impedance); the power associated to the generated radiation can be calculated by the average flux of the Poynting vector, the result is an emission at a frequency related to the Fourier component according to the following expression [15]:

$$A_{l,\lambda} = -\frac{Z_0}{2\beta_{gl}} \int_V \vec{J}_{Tl} \cdot \vec{E}_\lambda d^3x \quad => \quad P_{l,0,n} = \frac{\beta_{gl}}{2Z_0} \left| A_{l,0,n} \right|^2 \quad \text{where}$$

$$A_{l,0,n} = -\frac{Z_0}{\beta_{gl}} I_P \frac{C_l}{2} \frac{K}{\beta\gamma} F \frac{L}{\sqrt{ab}} \frac{\sin(\theta_l/2)}{\theta_l/2} i\, e^{i\theta_l/2} \qquad (5.19)$$

The expansion coefficients in Eq. 5.19 have been calculated using the modes of a metallic rectangular waveguide [6]. The spectral width of the single harmonic generated is related to the overall length of the macropulse, but also to the amplitude and phase stability of the bunches over the macropulse and thus is related to the quality of the RF source. All these considerations are useful to understand that there is a second degree of coherence in RF accelerator-based FEL operating at long wavelengths; this is related to the phase relation among all the bunches in the macropulse train that allows the generation of radiation as harmonics, of the fundamental RF. These harmonics can be spectrally very narrow.

After these considerations and wishing for the exploitation of all the features of a waveguide FEL, a compact undulator has been assembled inside the vacuum chamber of the facility described in the previous section [16]. The undulator is made of permanent magnets (SmCo) and is of a linear type, thus generates radiation that is linearly polarised in the horizontal plane. It has eight periods of 2.5 cm

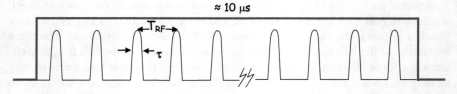

Fig. 5.10 The macropulse shape in a Radio-Frequency accelerator: τ is the microbunch length (see Fig. 5.6) and T_{RF} is the RF period

of length each for a total length of 20 cm. The undulator parameter K reaches the value of one at a gap of 6 mm corresponding to 6.3 KGauss of on-axis magnetic field. The waveguide used is a commercial one, (WR42), and it is copper made with an excellent surface quality and a vertical gap of $b = 4.318$ mm, the right value for the "zero slippage" condition (see below). The optical cavity has been realised by means of metal meshes and wire grid polarisers that can be considered as electron transparent mirrors; the use of these elements allowed to realise a straight-line e-beam propagation. An 80 μm wire spacing is used for the upstream mirror, resulting in a reflectivity higher than 99 % at wavelengths longer than 2 mm. This mirror intercepts the e-beam with an area transparency of about 90 %. The scattering of the residual 10 % does not affect significantly the e-beam qualities and the expected FEL performances. A wider wire spacing is used for the output coupler at the downstream end of the undulator providing a transmission of up to 15 %. In order to operate in the regime described by the synchronism (see Eqs. 5.4), the energy of the microtron accelerator has been lowered down to the value of 2.3 MeV [16].

The numbers chosen for the experiment exploit the potential expressed by Eqs. (5.4) and related to the characteristic of the waveguide operation in an FEL. With a suitable choice of the waveguide and electron-beam parameters, as described in Sect. 5.2.1) and analysing Fig. 5.3, the Δ parameter assumes a value such as the two solutions can merge into each other, generating a single broadband gain curve(see Fig. 5.11). Moreover, a peak of the emission corresponds to a frequency that has a group velocity $\beta_g = \partial(\omega/c)/\partial k$ equal to the electron beam velocity β_z. This condition corresponds to the so-called "Zero Slippage" condition that allows the electrons and the radiation to travel at the same speed inside the interaction region, maximising the energy exchange.

Observing Fig. 5.11 it can be seen that the bandwidth is no longer proportional to $1/N$ as indicated by Eq. (5.16), but to $1/\sqrt{N}$ with a wide flat top due to the sum of the two bandwidths of the two solutions. It is also worth to notice that, due to this wide band, the waveguide FEL can work also as a wide-band amplifier for frequencies close to the "zero slippage" condition. The frequency dependence of the gain, sketched in Fig. 5.11 presents the gain amplitude g_0 as a function of the frequency ν for the parameters of the ENEA Compact FEL as described above.

The experimental setup is visible in Fig. 5.12; two electron targets are used to monitor the e-beam position and shape, watching the fluorescence from the targets by means of two TV cameras. An electroformed horn transforms the rectangular guided transverse mode into a circular free space one and a 2000 line/inch copper mesh couples out the radiation vertically and transport it through a light pipe into the control room.

The radiation has been analysed outside the vacuum chamber and, regarding the spectral properties of this device, we also studied the cavity length tuning curve obtained by scanning the upstream cavity mirror with a remotely controlled stepping motor. A typical example of a cavity length tuning is reported in Fig. 5.13; from its observation, we may notice its complicated structure is extremely reproducible. This structure is strictly related to the presence of the harmonics of the accelerating RF that have to match with the longitudinal modes of the optical cavity. The importance of the

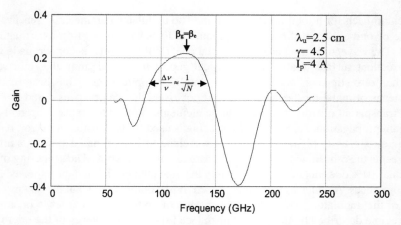

Fig. 5.11 The gain function at zero slippage for the ENEA millimetre wave FEL

Fig. 5.12 The Compact
FEL vacuum chamber with
the waveguide undulator-
resonator assembly

harmonics and their meaning has been described previously, but their presence has
been experimentally observed as can be seen from the two Fabry-Perot interferograms
reported in Fig. 5.14. Due to the dispersive properties of the waveguide, the matching
between the RF harmonics set and the cavity modes set is different at different
frequencies and thus the first interferogram, that has been taken at a cavity length
indicated by the first stylish arrow in Fig. 5.13, presents four harmonics, from the
37th to the 40th, of the RF giving a total bandwidth of about 5 % that is centred at a
wavelength of 2.6 mm. The second spectrum has been taken at a position indicated
by the second solid arrow, but it consists of a single harmonic to which corresponds
a bandwidth of less than 1 %. As can be seen also the emitted wavelength is different
and is 2.1 mm [16].

A proper reading of Figs. 5.13 and 5.14 tells us that a "zero slippage" waveguide
FEL has its own tuning mechanism, related to the dispersive properties of the guiding

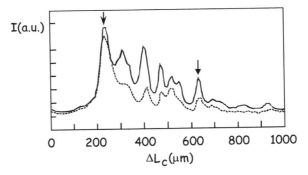

Fig. 5.13 The experimental cavity length tuning curve for the ENEA Compact FEL

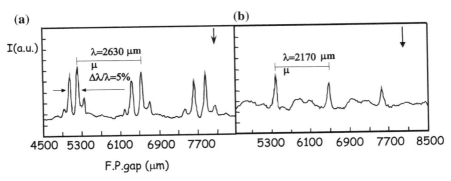

Fig. 5.14 The Compact FEL experimental Fabry-Perot interferograms: **a** taken at a cavity length corresponding to the stylish arrow of Fig. 5.13; **b** taken at a cavity length corresponding to the solid arrow of Fig. 5.13

structure, that involves only a cavity length tuning operation (very easy to realise), maintaining all the other parameters fixed; this is very important considering how difficult it is to find and hold the particle accelerators' working points.

As mentioned before, the spectral relative bandwidth of a single harmonic is related to the amplitude and phase stability of the RF source in the macropulse time scale and thus, in principle, can be very small. A very precise measurement can be done by means of a Fabry-Perot interferometer (FP) with a high finesse number. The FP that has been realised utilises high reflectivity metal meshes mirrors with which a finesse of 600 has been obtained; a corresponding resolution of $(\Delta\omega/\omega)_{RES} = 2.5 \times 10^{-4}$ has been achieved (see Fig. 5.15). The experimental results obtained with a gap of about 8 mm give a relative bandwidth of $(\Delta\omega/\omega) = 6 \times 10^{-4}$. This good spectral purity of the single harmonic comes directly from the spectral purity of the RF source accelerating the electron beam that for the ENEA Compact FEL is 3 GHz, and this is the reason why the distance between subsequent harmonics is 3 GHz as can be seen looking at Figs. 5.14 and 5.15.

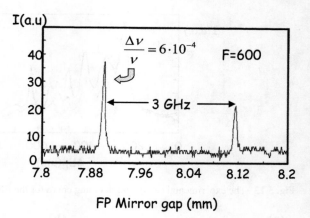

Fig. 5.15 A Fabry-Perot interferogram for the spectral line-width measurement

The performance of this source (about 1.5 kW of power over a macropulse duration of 4 μs corresponding to a peak power of about 10 kW over a micropulse duration of about 50 ps and its long time reliability make this source a unique tool for applications in many fields, from biology to solid state physics and cultural heritage conservation, where high peak powers but low average powers are required as in important issue.

5.4.3 The Smith-Purcell Grating Experiment

The Smith-Purcell FEL is the last experimental device realised at the ENEA Frascati FEL Facility. This has been made as a wide collaboration among ENEA, the University of Oxford, the Dartmouth College and the University of Essex [17]. The operating principle is simple; when electrons move along a metal grating an image charge is induced on the other side of the metal surface. The image charge will be generated at a distance related to the electron surface distance. When the electron moves, the overall effect is similar to an oscillating dipole and this phenomenon produces radiation emission.

A rigorous theoretical explanation of the Smith-Purcell mechanism goes beyond the purpose of the present work, but it can be said that the mathematical techniques are similar to that developed for the Cerenkov device. But now in addition, we have the relativistic contraction and thus the Doppler shift acts in a different way at different observation angles and the result is a radiation emission at wavelengths that depend on the angle θ (see Fig. 5.16); the synchronism condition can be summarised by a simple expression showing that the emission wavelength is proportional to the grating period l, to the inverse of the electron speed, expressed in terms of $\beta = v/c$, and also on the observation angle θ relative to the electron's velocity and to n that is the order mode of the radiation.

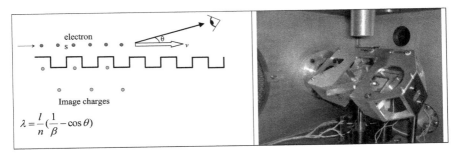

Fig. 5.16 The Smith-Purcell emission scheme and experimental layout

A proper optical system has been designed and realised to collect the emitted radiation and transport it to the detector. It has been designed for easy variation of the angle of observation (i.e. wavelength) and for an efficient collection of light over a wide range of emission angles. It consists of three gold-plated copper mirrors, held together in a rigid frame and rotating as a unit around a y-axis, through the centre of the grating. The range of angles accessible by the unit lies between 30° and 170°. The picture of the collecting mirror device and of grating realised for the experiment is visible in Fig. 5.16 [17].

The power of the emission as a function of the angle has been recorded and in comparison with the theory shows good agreement (see Fig. 5.17). It is worth to underline the extremely wide range of emission from about 400 μm to 4 mm associated with the variation of the angle θ; but in addition with these results obtained

Fig. 5.17 Experimental data points, fitted with various longitudinal profiles, all containing 85 % of the bunch particles inside 14 ps. 1: Gaussian; 2: Triangular with asymmetry factor $\varepsilon = 1.1$ and 3: Exponential

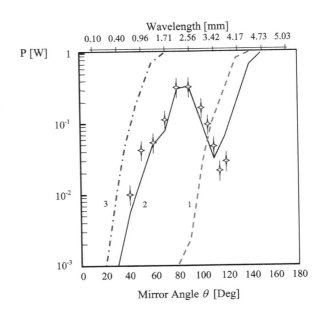

from the experiment, the demonstration of the linkage among the wavelength of the emitted radiation, its power and the electron pulse temporal profile. The graph reported in Fig. 5.17 indicates that the experimental data compared with simulations, well fit with the one using a triangular profile in a time duration of 14 ps, but it does not with Gaussian or exponential electron profile it is used instead; all simulations were performed with longitudinal profiles, all containing 85 % of the bunch particles. This techniques is an interesting and not destructive method for measuring the electron pulse profile in the picosecond time range.

5.5 The Compact Advanced THz Source

Describing the ENEA compact FEL experiment in the previous section, the problem in the presence of the RF harmonics in the generated radiation raised up. This experimental evidence is related to the fact that, as previously reported, a RF accelerator generates a train of bunches (microbunches) that is usually as long as several microseconds (see Fig. 5.10).

The expansion coefficients are reported in Eq. 5.19 and have been calculated using the modes of a metallic rectangular waveguide and finally, the power spectrum was calculated using the Poynting vector theorem.

But if the last expression of Eq. 5.19 is deeply analysed, it becomes evident that there is a phase term that can control the power emitting process. If, in fact, each electron bunch is treated as a collection of N_e particles distributed in the phase space (ψ, γ), each with its energy γ_j and position or phase $\psi_j = \omega_{RF} t_j$ along the longitudinal coordinate, the total radiated field will be simply calculated as the sum over the particles distribution. The following expression for the expansion coefficients can be then achieved [18]:

$$A_{l,0,n} = -\frac{Z_0}{\beta_{gl}} \frac{q}{T_{RF}} \frac{KL}{\sqrt{ab}} F \sum_{j=1}^{N_e} \frac{1}{\beta_{zj} \gamma_j} \frac{\sin(\theta_{lj}/2)}{\theta_{lj}/2} i e^{i\left(\frac{\theta_{lj}}{2} + l\psi_j\right)} \tag{5.20}$$

Now quite explicitly, a phase factor takes place that gives an interference mechanism. To maximise the extracted power it may be minimised the negative interference and this can happens when the particles are distributed in the phase space as close as possible to an ideal phase matching curve obtained keeping the phase term $\left(\frac{\theta_{lj}}{2} + l\psi_j\right) = \phi_l$ where the constant only depends by the harmonic number.

From these simple considerations, it can be understood that a further degree of coherence could be exploited to increase the performances of FELs operating at long wavelengths. This is the energy–phase correlation among the electrons distributed in the bunch. If the electrons might be distributed along the ideal curve in the phase space it could be obtained that any electron can emit in phase with the previous one

Fig. 5.18 The CATS experimental device: **a** the Linac and PMD assembly with the RF powering system; **b** the overall experimental assembly with the pumping system and the undulator vacuum chamber

and with the following one. This process can maximise the radiation emission up to saturation levels.

In order to exploit these features in ENEA, we wondered how could we manipulate the electron beam in order to obtain a distribution, in the phase-space, close to the ideal one. The answer that has been found by the ENEA team is a RF device downstream the proper electron beam accelerator and, within this framework, it has been realised a new compact device that we have called Compact Advanced THz Sources (CATS) [19]. This FEL present the new element, that we called "Phase Matching Device" (PMD), that manipulates the electron beam in the phase space. In Fig. 5.18a), it is illustrated that the accelerating system that is made up by a β-graded S-band Linac accelerator, with a pulsed triode gun, that accelerates an electron macropulse current of about 250 mA to a kinetic energy of 2.3 MeV; the corresponding peak current is of about 5 A. Downstream from the Linac end we have the PMD RF section; the drift space and the RF phase shift of the PMD must be set to have the reference electron passing through the centre of the PMD with a phase close to zero. In the PMD device, we can change the RF field amplitude and phase relatively to the Linac, in so doing the higher energy electrons, arriving first at the PMD, are decelerated, while the lower energy electrons are accelerated and emerge first from the PMD. The result is exactly a rotation of the electron distribution in the phase space.

The picture illustrating the complete setup is Fig. 5.18b where the RF system, including a variable attenuator and a phase shifter for the appropriate manipulation control, is partly visible. After a conventional straight line transport channel, we have the vacuum chamber hosting a compact undulator of 16 periods of 2.5 cm; the magnets are of NdFeB and are arranged in the Halbach configuration. The gap is about 10 mm capable to host a waveguide having a vertical internal dimension of

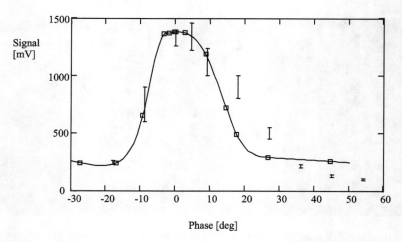

Fig. 5.19 The CATS total emitted power measured as a function of the RF relative phase between LINAC and PMD (*bars*) compared with simulation (*squares and spline*)

6 mm and the corresponding K value is 0.75. The total length of the CATS FEL source, including, the beam dump, is of the order of 2.5 m [19].

Two important experimental results can be illustrated: The first one is shown in Fig. 5.19 where it is reported the total power recorded by a pyroelectric detector (vertical bars) measured as a function of the RF relative phase between LINAC and PMD. Squares boxes show the expected power at selected values of the phase according to the theory [18]. The continuous line is the spline interpolation of the calculated data. When the phase difference is equal to zero the source emits at 760 μm. All these measurements have been carried out with a RF electric field amplitude, on the PMD equal to one half of that on the Linac [19].

The second experimental result is shown in Fig. 5.20 and is related to the radiation spectral analysis by means of Fabry-Perot interferograms recorded with at least two interferometric orders, for four different wavelengths emitted. The data are recorded at different RF phase and the different wavelengths come from the fact that the PMD become also an accelerating, or decelerating, structure when changing the relative RF phase. It is important to underline how wide is the tunability range, almost one octave, making this source an appealing one for spectroscopic measurements. The ordinate axis reports the pyroelectric detector signal and the relative amplitude differences are real. A power estimation has been done for the stronger line at 750 μm, giving a result higher than 1 kW over about 6 μs of macropulse length [19].

It is important to stress that the reported power level has been obtained after a single pass of the electron through the undulator without any optical cavity. The CATS device, in fact, does not need to store the radiation in the interaction region to produce stimulated mission, because the intrinsic coherence of the electron beam and the proper correlation of the electrons emissions, induced by the PMD do the

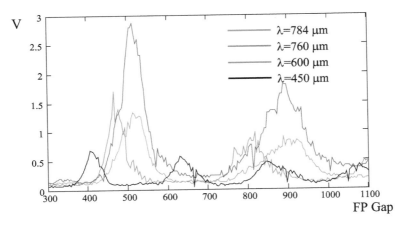

Fig. 5.20 The Fabry-Perot spectra recorded at different values of the RF phase difference

same work of the bunching process, but the spectral bandwidth is not affected now by the presence of an optical cavity that act as a spectral filter.

The last important thing to be noticed looking at Fig. 5.20 is that due to this large bandwidth, the spectrum of the CATS device, as for the ENEA compact FEL, is made by the composition of single frequency components that can be easily extracted by means of optical systems. This is a very important peculiarity of this device respect to conventional FEL sources, because now high-resolution spectroscopy can be performed, exploiting the individual frequency components, and also with an appreciable power for many of the harmonics composing the spectrum. Moreover, another interesting result is that the single frequency, being an harmonic of the RF has a temporal structure equal to that of the RF macropulse. If we look at the whole bandwidth, the temporal structure is the well-known train of microbunches separated by the RF period. In conclusion, FEL can be considered as a convenient, flexible and powerful source for the generation of coherent radiation in the THz spectral region.

References

1. A. Gover, A. Yariv, Collective and single electron interactions with electromagnetic waves in periodic structures and free electron lasers. Appl. Phys. **16**, 121 (1978)
2. R.H. Pantell, in *Energy and Momentum Conservation Requirements for Electron Interactions with Electromagnetic Radiation*, ed. by S.F. Jacobs. Free Electron Generators of Coherent Radiation, Physics of Quantum Electronics, vol. VII (Addison Wesley Publ. Co. 1980)
3. G. Dattoli, A. Renieri, A. Torre, *Lectures on the free electron laser theory and related topics* (World Scientific Publishing Co. Pte, Ltd, 1993)
4. J.D. Jackson, *Classical Electrodynamics* 3rd edn., Sect. 14.7 (John Wiley and Sons Inc., 1999)
5. P. Luchini, H. Motz, *Undulators and Free-electron Lasers* (Oxford University Press, 1990)
6. A. Doria, G.P. Gallerano, A. Renieri, Kinematic and dynamic properties of a waveguide FEL. Opt. Commun. **80**, 417 (1991)

7. G. Dattoli, A. Renieri, in *Experimental and Theoretical aspects of Free Electron Lasers*, ed. by M.L. Stitch, M.S. Bass, Laser Handbook, vol 4 (Elsevier, North Holland, 1983)
8. J.M.J. Madey, Relationship between Mean Radiated Energy, Mean Squared Radiated Energy and Spontaneous Power Spectrum in a Power Series Expansion of the Equation of Motion in a Free Electron Laser. Nuovo Cimento **50B**, 64 (1979)
9. H. Wiedemann *Particle Acceleratori Physics* (Springer-Verlag, New York, 1995)
10. G.P. Gallerano, A. Doria, E. Giovenale, A. Renieri, Compact free electron lasers: from Cerenkov to waveguide free electron lasers. Infrared Phys. Technol. **40**, 161 (1999)
11. F. Ciocci, G. Dattoli, A. Doria, G.P. Gallerano, G. Schettini, A. Torre, Spontaneous emission in a double dielectric slab for Cerenkov Free Electron Laser operation. Phys. Rev. A **36**, 207 (1987)
12. F. Ciocci, G. Dattoli, A. Doria, G. Schettini, A. Torre, J.E. Walsh, Spontaneous emission in Cerenkov FEL devices: a preliminary theoretical analysis. Il Nuovo Cimento D **10**, 1 (1988)
13. F. Ciocci, A. Doria, G.P. Gallerano, I. Giabbai, M.F. Kimmitt, G. Messina, A. Renieri, J.E. Walsh, Observation of coherent millimetre and submillimetre emission from a microtron driven Cerenkov Free Electron Laser. Phys Rev. Lett. **66**, 699 (1991)
14. J.S. Nodvick, D.S. Saxon, Suppression of Coherent Radiation by Electrons in a Synchrotron. Phys. Rev. **96**, 180 (1954)
15. A. Doria, R. Bartolini, J. Feinstein, G.P. Gallerano, R.H. Pantell, Coherent emission and gain from a bunched electron beam. IEEE J. Quantum Electr. **29**, 1428 (1993)
16. F. Ciocci, R. Bartolini, A. Doria, G.P. Gallerano, E. Giovenale, M.F. Kimmitt, G. Messina, A. Renieri, Operation of a compact Free Electron Laser in the millimetre-wave region with a bunched electron beam. Phys. Rev. Lett. **70**, 929 (1993)
17. G. Doucas, M.F. Kimmitt, A. Doria, G.P. Gallerano, E. Giovenale, G. Messina, H.L. Andrews, J.H. Brownell, Determination of longitudinal bunch shape by means of coherent Smith-Purcell radiation. Phys. Rev. Special Topic Acc. Beams **5**, 072802 (2002)
18. A. Doria, G.P. Gallerano, E. Giovenale, S. Letardi, G. Messina, C. Ronsivalle, Enhancement of coherent emission by energy-phase correlation in a bunched electron beam. Phys. Rev. Lett. **80**, 2841 (1998)
19. A. Doria, G.P. Gallerano, E. Giovenale, G. Messina, I. Spassovsky, Enhanced coherent emission of terahertz radiation by energy-phase correlation in a bunched electron beam. Phys. Rev. Lett. **93**, 264801 (2004)

Chapter 6
Physics and Applications of T-Rays

Graeme P. A. Malcolm, David A. Walsh and Marc Chateauneuf

Abstract This chapter introduces the concept of using conventional optical/infrared lasers to produce terahertz radiation. These techniques are of great interest not only due to the high spectral, spatial, and temporal qualities of the terahertz radiation that they can produce, but also due to the practicality of compact, room temperature systems that are based on turn-key, off-the-shelf technologies. More specifically, this chapter describes the intracavity terahertz parametric oscillator (ICTPO), which is one of the simplest and most practical laser-based terahertz sources. In the first part the considerations necessary when designing an ICTPO are discussed, and the description of a realised system is given. The ICTPO produced nanosecond terahertz pulses that were tunable from <0.9 to >3.05 THz, with peak powers of several Watts. Typical terahertz bandwidths were only ~50 GHz, and were further reduced to 500 MHz with the simple addition of etalons in the optical fields. This narrow bandwidth makes the ICTPO complementary to terahertz time domain spectroscopy, where the bandwidth spans several to tens of THz. In the second part the application of the ICTPO in standoff spectroscopy of explosives is discussed. Here, the narrow linewidth allows the available power to be tuned into atmospheric transmission windows, and the non-reliance on coherent detection reduces artificial features in the obtained spectrum. An example is given where the ICTPO was used for real-world standoff spectroscopy at a range of 8 m (16 m round-trip), yielding one to two orders of magnitude improvement in the useable range compared to other sources.

G. P. A. Malcolm (✉)
M Squared Lasers Ltd., Venture Building, 1 Kelvin Campus, West of Scotland Science Park, Glasgow G200 SP, UK
e-mail: Graeme.malcolm@m2lasers.com

D. A. Walsh
School of Physics & Astronomy, University of St Andrews, North Haugh, St Andrews KY169 SS, UK

M. Chateauneuf
Defence R&D Canada Valcartier, 2459 Boul. Pie XI North, Québec, QC G3J1X5, Canada

M. Perenzoni and D. J. Paul (eds.), *Physics and Applications of Terahertz Radiation*, Springer Series in Optical Sciences 173, DOI: 10.1007/978-94-007-3837-9_6,
© Springer Science+Business Media Dordrecht 2014

149

Keywords Terahertz source · Intracavity · Optical parametric oscillator · OPO · ICTPO · Standoff spectroscopy · Phase matching · Lithium niobate

6.1 Intra-Cavity THz Generation in OPOs

6.2 Introduction

In selecting a method of terahertz generation it is not only important to consider the characteristics of the radiation required, but also the practicality of the source. If the source requires cryogenic cooling this adds significantly to its complexity and reduces its practicality significantly. For example, an ongoing maintenance schedule for the replenishment of liquid nitrogen and/or liquid helium could be required, and vacuum systems may be needed to for insulation. More severe practicality issues arise in the case of the free electron laser (FEL), where although the terahertz output is powerful, tunable and has a good spatial mode, the FEL itself is typically a very large and costly multiuser installation to which access may not be available, or possible, for the application in mind. To give a couple of examples, it would be no simple matter (and probably not cost effective!) to integrate an FEL into the quality control line at a factory, or to perform atmospheric measurements at remote locations. It is partially due to these considerations that it has been the development of a number of conventional laser-based sources which can easily be made into convenient, turn-key bench-top systems that has enabled and invigorated scientific research and applications in the terahertz frequency range. A prime example of this is the use of ultrashort pulse Ti:Sapphire lasers to generate broadband terahertz pulses via either nonlinear optics or photoconductive switches (terahertz time domain spectroscopy, or THz TDS, systems), which has fuelled a vast amount of research since its inception. THz TDS is covered later in Chap. 8 of this book. One of the main limitations of THz TDS is the necessity of using Fourier techniques to obtain spectral information; care must be taken to ensure that the system is set up correctly (both alignment and reflections must be considered) to avoid distortions in the retrieved spectrum, and it can take minutes to acquire enough data for a single measurement.

There are a number of schemes for the generation of THz radiation using standard laser sources, most of which are based on second-order nonlinear optical processes [1]. Conceptually the simplest is difference frequency generation (DFG). Here, two laser-like beams are overlapped in a nonlinear medium, and the beat (difference) frequency between them is generated through the interaction with a nonlinear polarisation response. The tuning of either one or both of the sources allows the generation of tunable terahertz output, and has been demonstrated in a variety of configurations mainly using pulsed lasers as the nonlinear process requires high intensity to operate efficiently [2–5]. However, this method requires two sources to be synchronised, and efficient generation requires careful alignment of the beams to achieve optimal mode- and phase-matching (which may change with frequency tuning), all of which adds complexity to the source. The aforementioned technique of THz TDS has an advantage over DFG in that only one laser beam is required. In these systems the generation process can be thought of as a special case of DFG, where the frequency mixing occurs between the spectral components of a single pulse. Therefore, given a laser pulse of sufficient bandwidth, such as the output of a modelocked Ti:Sapphire laser, pulses of radiation at THz frequencies can be produced. The drawbacks are that the broadband THz output of a TDS system requires Fourier methods to retrieve spectral information, and that modelocked lasers are typically more complex than their Q-switched relatives. There is, fortunately, a way that terahertz radiation can be generated from a single, nanosecond pulsed, input beam. In the DFG process both of the lower frequency waves experience gain at the expense of power in the highest frequency "pump" wave. If the beam is of sufficient intensity, then the second beam required for DFG experiences a gain sufficiently high that it can build up from noise photons, rather than being injected from another source. This type of source is known as an optical parametric generator, or if the generated wave is resonated in an optical cavity to enhance the gain (similarly to a laser cavity) it is known as an optical parametric oscillator (OPO) or THz parametric oscillator (TPO) if the desired output is at THz frequencies. As the source now "self seeds" no second laser-like source needs to be synchronised and aligned, and a very simple device can be realised.

Generation of THz radiation through parametric down conversion was, arguably, first demonstrated in 1969 by Gelbwachs et al. [6]. Although in this work the THz radiation was not usefully extracted from the device, its presence can be inferred from the near infrared radiation that was observed. The first time the THz radiation was extracted was by the same research group later that year [7] and work progressed until roughly 1975 when Piestrup et al. [8] demonstrated a continuously tunable, nanosecond pulse Q-switched laser pumped, system that set out the basic design still used today. There was then a gap of around two decades where no further work on the techniques were published until Kawase et al. in 1996 [9], who then proceeded to make many improvements to the technology including improved THz output coupling techniques [10] and narrower linewidths [11]. All of these developments, however, did not improve on the requirement of relatively powerful nanosecond pulse Q-switched lasers needed to pump the devices above threshold.

The TPO is a singly resonant oscillator (SRO), i.e., one of the down converted waves is resonated to provide multiple passes of the nonlinear crystal in order to provide more gain. A well-known technique to lower device thresholds further is to resonate both generated waves (known as the doubly resonant oscillator, or DRO), which can lower the threshold by orders of magnitude at the expense of complexity. The generated waves must then satisfy the conservation of energy and also match allowed modes of the resonators used, and so ease of tunability and stability are sacrificed. The use of DRO is also not a practical option in the TPO as the THz waves are highly absorbed in lithium niobate [12] (the most practical and reported nonlinear crystal used in TPOs) such that a round trip of the THz wave cavity would result in a net loss rather than gain. Another well-known method to reduce laser threshold is to increase the pump field intensity by resonating the pump wave itself. This technique can be very effective as the enhancement can be more than an order of magnitude, but there are significant difficulties in that the resonant cavity must be mode matched with, and locked to the frequency of the pump wave. There are a number of techniques (e.g., Pound–Drever–Hall cavity locking [13]) that can achieve this locking at the cost of added complexity and expense (note that these techniques also require a single frequency pump source), but most practical techniques require a continuous wave pump to prevent the cavity drifting out of resonance. For a thorough analytical discussion of the techniques mentioned here and the physical processes behind parametric generation methods the reader should refer to Ebrahimzadeh and Dunn [14].

A very successful method to reduce the power requirements has been pioneered by work at the University of St Andrews and is that of the intra-cavity TPO (ICTPO), where the TPO is placed inside the resonant cavity of the pump laser and thereby directly accesses the intense intra-cavity optical field. This method obviates the need for cavity locking techniques and mode matching optics and produces devices that are more robust and compact. Additionally parasitic losses, such as Fresnel reflections, diffraction, laser output coupling efficiency, etc., that would be incurred in pumping an extra-cavity system are minimised. This technique was first reported by Edwards et al. in 2006 [15] where the great improvement in power requirements was realised, and has since been developed into a practical commercial system.

Of the many techniques that are used to generate THz radiation, the parametric oscillator has many advantages. The radiation produced can have excellent, laser-like, beam quality which is important in both imaging and stand-off detection techniques, and also exhibits a high spectral brightness combined with a high degree of continuous tunability, making it a powerful spectroscopic tool. In addition the devices themselves operate at room temperature, and are practical table-top systems that do not necessarily have high power requirements. The potential of parametric generation in the visible and infrared regions has already proven its worth, and for these reasons there has been much development of the technique over the past years as a method to fill the so-called 'THz gap'. In this section the design and characteristics of the ICTPO first reported by Edwards et al. [15] and a simple yet highly effective method for the reduction of the THz output's bandwidth [16] will be described.

6.3 The Nonlinear Medium, Phase Matching and Tuning

The TPO is a special case of an OPO. Traditionally, the OPO utilises a second order nonlinear response in a medium to down convert (or 'split') 'pump' photons into two new, lower frequencies that lie within a single, to a few, octaves of one another. The split of the pump photon into the two lower energy photons has to satisfy the conservation of energy, in that their energies summed equal that of the pump photon. In the case of the TPO, one of the down converted photons is very close in frequency to the pump photon. This photon is designated as the 'signal' photon. The other generated photons corresponds to the difference in the energies of the pump and signal photons and are therefore being pushed down into the THz frequency range (i.e., around 100 times lower in frequency for a NIR pump!). This is a special case for a number of reasons:

1. The nonlinear material has to be transparent at all three frequencies—and not many materials are transparent both at convenient laser wavelengths and also at THz frequencies.
2. The small signal gain coefficient in an OPO/TPO is inversely proportional to $\sqrt{\lambda_i \lambda_T}$—in the THz region λ_T is of the order of 100 times greater than the 'pump' and 'signal' waves so it is clear that the gain is significantly reduced.
3. The condition of phase matching must be possible for the conversion process to experience an overall net gain (i.e., the generated waves must remain in constructive phase with previously generated waves).

The majority of ICTPOs to date have been based around lithium niobate ($LiNbO_3$) crystals, due to two of the material's properties. First, the nonlinear coefficient, which is already large, is enhanced due to an interaction with the A1-symmetry soft phonon mode (a "polariton resonance" as it couples with electromagnetic waves) at $\sim$248 cm^{-1} ($\sim$7.4 THz). Unfortunately the existence of this polariton resonance is a double edged sword. While its presence greatly enhances the nonlinear process, making the crystal a feasible choice, it also means that the generated THz radiation couples easily to the phonon modes of the crystal lattice and thus is readily absorbed. Absorption coefficients between 20 and 80 cm^{-1} in the range 1.2–2.4 THz have been measured [17]. To minimise the problem, the THz wave must exit the crystal with a path length that is as short as possible. Second, the dispersion in the material enforces a noncollinear phase matching geometry (see Fig. 6.1) that, although reducing the nonlinear gain due to walk-off between the three waves, provides a simple tuning mechanism and more significantly allows the THz wave to be generated at a large angle to the pump wave. This, in conjunction with positioning the generation region at the side of the crystal, allows the THz wave to traverse a short path within the crystal and so experience a much smaller absorption.

When pumped with high fluence Q-switched lasers, it is quite easy to reach the damage threshold of lithium niobate and this is a limiting factor in the maximum power of the ICTPO. A common method to increase the damage threshold of the material is to dope it with MgO whilst it is being grown [18]. Such crystals have been tested for use in the TPO where a level of 5 % MgO doping proved to raise the damage threshold and have an enhanced efficiency for THz generation [19].

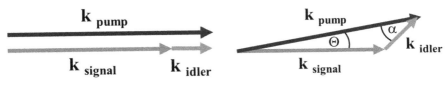

Fig. 6.1 Vector diagrams illustrating phase matching solutions: *left* the conventional collinear phase matching solution; *right* the noncollinear phase matching condition existent in lithium niobate-based ICTPOs. N.B. k vectors are not drawn to scale to enhance clarity

Fig. 6.2 A graph showing the dispersion of the k vector directions as defined in Fig. 6.1, all angles are defined for the waves within the crystal

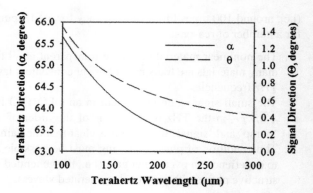

With knowledge of the refractive index at infrared [20] and THz [12] wavelengths the phase matching solutions can be calculated over the 1–3 THz range. The angles of the signal and THz waves with respect to the pump wave for these solutions are shown in Fig. 6.2. It can be seen that the signal angle varies with the THz wavelength generated, therefore by defining the signal angle the (IC)TPO can be tuned. As the (IC)TPO employs a resonator to provide multiple passes of the nonlinear crystal to effect a higher gain, setting the angle of the resonator to match the desired wavelength ensures that only this solution experiences sufficient gain to operate.

The propagation angle of the THz radiation and the high refractive index of lithium niobate ($\sim$5.2) result in total internal reflection at the crystal-air interface. As a result, it is not possible to extract THz radiation in this way as the internal reflection would result in all of the THz radiation being absorbed in the lithium niobate crystal. The application of silicon prisms overcomes this problem. In the first instance the total internal reflection angle is reduced due to a different combination of refractive indices at the crystal interface enabling the THz wave to escape into the silicon. In addition, the prism is cut such that the wave is normal to the surface on exit [10]. Other methods have been tested, including cutting the crystal corner such that the THz hits normal to the surface [7], and cutting a grating coupler into the crystal side [9]. The downside of this technique is that the angular dispersion of the THz wave as it is tuned is quite large compared to the silicon prism technique, a property that is a disadvantage for most applications. The angular dispersion has been calculated for the corner cut (CC), grating coupler (GC), and silicon prism(SPA) output coupling techniques and is shown in Fig. 6.3, and illustrates why the silicon prism array was selected for the ICTPO.

6.4 System Design

The design of the ICTPO was based around a simple electro optically Q-switched Nd:YAG laser—a robust laser system with excellent reliability. Nd:YAG lasers are widely employed and so laser components such as pump laser diodes, mirrors and

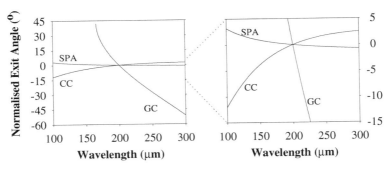

Fig. 6.3 Graphs showing the calculated angular dispersion of the output coupled THz wave for CC, GC and SPA approaches. All plots are normalised to the exit angle of 200 μm output and the grating design is that reported in Ref. [9]

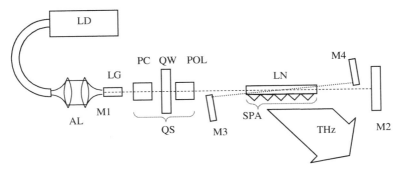

Fig. 6.4 The experimental design of the intersecting cavity THz OPO. The *dashed* and *dotted lines* represent the Q-switched laser and signal waves, respectively

Q-switching optics are readily available making for a very practical basis for the ICTPO. The experimental set-up of the ICTPO is shown in Fig. 6.4. Referring to Fig. 6.4, the Q-switched laser cavity was formed by mirrors M1 (R > 98 % @ 1,064 nm, high transmission @ 808 nm) which was coated directly on the Nd:YAG laser crystal and M2 (R = 90 % @ 1,064 nm) which was a standard dielectric laser mirror. The separation of the mirrors was ~37 cm. The output coupler was chosen for the ease with which intracavity pulse intensity could be calculated from the laser output. Even higher intracavity intensities and so lower thresholds would result from using a reflectivity of 100 % for M2, with the output coupling of the entire laser field (excluding that lost by parasitic mechanisms) into the down converted waves; however the lower reflectivity was chosen to ease device characterisation. Components QW (quarter wave plate), PC (Pockels cell) and POL (air-spaced cube polariser) constitute a quarter-wave electro-optic Q-switch (QS) and set the polarisation of the laser to be orthogonal to the plane of the diagram. The laser gain medium (LG) was a cylindrical Nd:YAG crystal (1.3 % doped) 7 mm long and 4 mm in diameter, on the end of which mirror M1 had been deposited. Pump light at 808 nm was provided by a temperature stabilised quasi-continuous wave laser diode module (LD) and was delivered via an optical fibre (800 µm core diameter) to a pair of aspheric lenses (AL), which image the end of the fibre into the LG with a 1:1 ratio. End pumping provides an efficient coupling of energy into the desired laser mode diameter of ∼1 mm and is much more efficient than side excitation schemes or flash lamp pumping [21].

The lithium niobate nonlinear crystal (LN) had the dimensions of $5 \times 5 \times 50$ mm and had its crystallographic z-axis parallel to the polarisation of the laser (i.e., also coming vertically out of the page). The x-axis of the crystal was parallel to the laser axis and the 50 mm long side of the lithium niobate. Each end was anti-reflection coated at 1,064 nm and the 5×50 mm side faces, through which the THz was extracted, were polished to optical specifications. Against one of these polished sides, an array of seven high resistivity (>10 kΩ·cm) silicon prisms was held in place by a specially designed mount to ensure good optical contact so that the THz radiation may be coupled out.

The ICTPO signal cavity was formed by the plane mirrors M3 (high reflectance for signal) and M4 (R $\sim 98\%$ for signal) and was 13 cm in length. Due to the high difference in frequency of the pump and THz waves, the signal wavelength is very close to that of the pump, enabling the use of standard Nd:YAG laser mirrors. Both of these mirrors were on individually adjustable mounts which were themselves mounted on a common swing arm that pivoted around the centre of the nonlinear crystal. It was by rotating the angle of this "intersecting" cavity that the angle at which feedback is provided was changed, thereby selecting different phase matching solutions such that the signal and THz waves were tuned.

6.5 Performance and THz Characteristics

To characterise the down conversion performance of the ICTPO, the energy characteristic of the Q-switched laser was measured with the ICTPO running and also when the signal cavity was blocked. The results, plotted in Fig. 6.5, show that the laser output was heavily depleted by the operation of the ICTPO, showing the characteristic "clamping" of the laser power at down conversion threshold levels. It can also be seen from this plot that the pump wave was depleted by ∼50 % when running significantly above threshold.

Oscillation threshold was reached when the Nd:YAG was pumped with 10 mJ of primary pump light at 808 nm. At this level the laser emitted pulses were 1.8 mJ in total (parasitic laser cavity losses were accounted for) at 1,064 nm. Compared to the reported 18 mJ threshold of a similar extracavity device [10], the benefits of the intersecting cavity approach are clear, in this case reducing the laser pulse energy required by an order of magnitude.

When pumping at roughly twice the ICTPO threshold with a pump diode pulse energy of ∼19 mJ, the laser produced pulses at 1,064 nm of ∼3.6 mJ in total in the absence of ICTPO operation (achieved by physically blocking the OPO signal cavity), reducing to 1.8 mJ in total with the ICTPO down converting. This indicates

Fig. 6.5 The typical energy characteristic of the Nd:YAG laser with the ICTPO tuned to 1.6 THz. It can be seen that the ICTPO operates effectively, down converting ∼50 % of the pump light at twice threshold. The laser could be run at over twice lasing threshold implying that a large portion of the population inversion was being extracted by the QS pulse

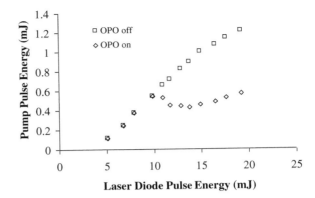

that a total of 1.8 mJ of signal and THz radiation was being generated. Unfortunately, just as in the case with extracavity devices, only a small fraction of this was converted into THz photons due to the large difference in frequency; the quantum defect is as low as 0.35 % at 1 THz and up to 1.1 % at 3 THz. Here, the ICTPO was generating ∼190 μm THz waves, indicating that an estimated 10.2 μJ of THz radiation was being produced per pulse. However, when using a calibrated composite silicon bolometer (QMC Instruments Ltd. model QSIB/2) to measure the pulse energy only ∼20 nJ was recorded. The primary cause of this ∼1,000-fold decrease in energy, which was observed using both intracavity and extracavity techniques, is believed to be the aforementioned absorption of THz frequencies in lithium niobate. In addition, a factor of two reduction can be attributed to THz being generated in the opposite direction on the backward pass of the pump and signal waves. In any case, this value exceeds significantly the THz pulse energy quoted for a similar device reported in [10] of 200 pJ (which required 30 mJ of 1,064 nm pump energy, rather than the 3.6 mJ of the ICTPO), and is more than adequate for many spectroscopic and imaging applications.

Figure 6.6 shows the temporal characteristics of the pump and signal pulses at twice threshold. Again, it can be seen that the pump undergoes ∼50 % down-conversion. Importantly, it can also be seen that the signal pulse builds up swiftly just after the peak of the pump pulse. These features indicate that near optimal extraction of the energy stored in the Nd:YAG crystal was achieved, followed by rapid cavity dumping of the oscillating pump field into THz and signal. The profile of the signal also provides some indication of the THz pulse duration, as it must be generated mainly during the build up phase of the signal wave. This suggests the THz pulse is <10 ns in duration, and for a THz pulse energy of around 10 nJ indicates a peak power of >1 W.

Fig. 6.6 The temporal profiles of the pump and signal
waves. The cavity dumping
of the pump wave is clearly
illustrated. The *dotted line*
represents the measured pump
profile in the absence of down-
conversion, again indicating a
~50 % depletion

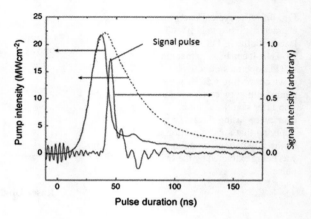

6.6 Gain and Threshold Calculations

In order to check the performance of the ICTPO the build up time of the signal wave
at the ICTPO threshold can be used to calculate the gain present, and compare it with
the gain theoretically calculated from to the pump intensity.

It is recalled that for a crystal length L the single pass multiplicative gain G is
expressed by:

$$G = \cosh^2{(\Gamma L)} \tag{6.1}$$

where Γ is the small signal gain coefficient, and is expressed by:

$$\Gamma^2 = \frac{8\pi^2 d_{\text{eff}}^2 I_p}{c\varepsilon_0 n_p n_s n_T \lambda_s \lambda_T} \tag{6.2}$$

Here, n and λ are the refractive index and wavelength at the wave denoted by the sub-
scripts p, s and T for the pump, signal and THz waves, respectively. ε_0 is the permittivity of free space, I_p is the intensity of the pump wave, and d_{eff} is the effective
nonlinear coefficient.

In this situation a modification has to be made in that the length L has to be
broken down into smaller segments over which the three waves interact (the effective
interaction length) because of the walk-off of the THz wave [22]. The total single
pass gain is then calculated by multiplying the gain of each of the segments

$$G = \cosh^{(2N)}{(\Gamma l)} \tag{6.3}$$

where l is the length of a segment, and $N = L/l$ is the number of segments that make
up the length of the crystal. This situation is shown in Fig. 6.7 for the case of the
walk-off of the THz wave.

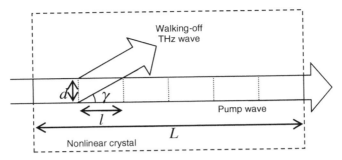

Fig. 6.7 An illustration of the reduction in interaction length due to the THz wave propagating at a large angle to the other waves

From the analysis put forward by Brosnan and Byer [23] it is generally accepted that around 140 dB of net gain is required for a pulsed OPO to reach threshold. That is to say that for M round trips of the signal cavity (within the temporal envelope of the pump pulse), the following expression must be satisfied

$$140\,\text{dB} = G^M = \left(\cosh^{(2N)}(\Gamma l) \right)^M = 10^{14} \tag{6.4}$$

Once rearranged, this can be used to estimate the required value of Γ.

The pump pulse duration (t_p) was ~50 ns at threshold, and given that the signal cavity was ~20 cm in optical length (L_{opt}) gives:

$$M = \frac{t_p}{\left(\frac{2L_{\text{opt}}}{c} \right)} \sim 37.5 \tag{6.5}$$

The result of Eq. 6.5 gives the number of round trips of the signal cavity during which parametric gain is present. Considering the interaction of the three waves to be limited by the walk-off gives an interaction length of: $l = \frac{1}{\tan 65°} \approx 0.5$ mm. In the intracavity regime the pump wave is resonant and so gain is present in both passes of the signal wave through the lithium niobate, doubling the crystal length per round trip. Hence the number of gain sections is: $N = \frac{2 \times 50\,\text{mm}}{0.5\,\text{mm}} = 200$. Again assuming that 140 dB of gain is required to bring a pulsed device to threshold from noise gives a required gain of:

$$\Gamma = \frac{\cosh^{-1}\left(2*M*N\sqrt{10^{14}} \right)}{l} \approx 1.3\,\text{cm}^{-1} \tag{6.6}$$

This result can now be compared to the gain calculated for the present intracavity intensity. The intracavity peak power is calculated from the pulse energy extracted from the output coupler ($PE = 0.6$ mJ), the reflectivity of the output coupler ($R_{oc} = 0.9$), and the pulse duration as ($t_p = 50\,\text{ns}$): $P = \frac{PE}{t_p} \times \frac{1}{1-R_{oc}} =$ 120,000 W. The peak intra-cavity pulse intensity is therefore $1.53 \times 10^{11}\,\text{W} \cdot \text{m}^{-1}$ for a beam diameter of 1 mm. If an enhanced d_{eff} of 125 pm $\cdot\text{V}^{-1}$ [10] is assumed then Eq. (6.2) can be used to calculate a gain coefficient of $\sim$1.2 cm^{-1}, which is in agreement with the estimated gain required.

6.7 THz Output Characteristics

The output of the ICTPO has been tuned from 900 GHz up to 3.05 THz (verified by measurements of the pump and signal with a wavemeter) with pulse energies circa 10 nJ up to $\sim$2.1 THz, beyond this the power dropped off rapidly as illustrated in Fig. 6.8. At the lower THz frequency end of the tuning range the gain was reduced due to the increasing separation of the THz wave frequency and the polariton resonance (reducing the enhancement of the nonlinear coefficient). The angle required between the signal and pump cavities was also a limiting factor. Frequencies lower than 1 THz require a signal cavity angle of <1°, and considering that the pump and signal

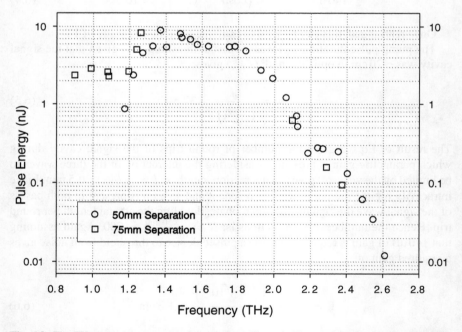

Fig. 6.8 The THz power spectra recorded using a calibrated silicon bolometer. Data is shown for two signal cavity configurations, with signal mirrors either 50 or 75 mm from the MgO : LiNbO$_3$ crystal, illustrating the extension of the tuning range. A number of dips are present in the data that correspond to the absorption lines of atmospheric water vapour present in the THz path [24]

beams were both ~1 mm in diameter a signal cavity 13 cm long would then clip the intra-cavity laser field. The signal cavity could be lengthened to ease this issue, but only to some extent: the signal cavity length was limited by the laser resonator cavity components, and excessive lengthening of the cavity would increase the build up time of the signal wave, degrading performance. Figure 6.8 shows the extension of the tuning range from 1.2 THz down to 900 MHz by the lengthening of the signal cavity by 25 mm at either end.

The generation of higher frequencies faced other problems: as the signal cavity angle increased the overlap of the pump and signal waves decreased, and the signal cavity began to be clipped by the crystal aperture. A compromise was found between moving the generation region away from the crystal face to avoid clipping, and keeping it close enough to minimise the significant absorption loss, which increases as the generated THz frequency approaches that of the polariton resonance.

With typical pulse energies of 10 nJ, it was possible to saturate the calibrated bolometer used for characterising the source. On the basis of averaging over 16 pulses the noise equivalent energy was ~0.3 pJ using this bolometer, giving a dynamic range in excess of 35,000:1. The THz output was also successfully measured using relatively inexpensive, room temperature detectors—the Golay cell (Tydex GC-1P) and pyroelectric (Spectrum Detector SPH-62 THz) detectors, with minimal detection energies of 4 and 100 pJ, and dynamic ranges of 2,500:1 and 100:1, respectively. Without doubt the application of phase sensitive detection schemes would improve these dynamic ranges even further.

6.8 Terahertz Beam Quality

On exiting the silicon prism array the THz wave diverged rapidly in the vertical plane, as the wavelength is large in comparison to the height of the generation region that was used (the pump beam profile, i.e., ~1 mm). However, in the horizontal direction the THz wave was generated over several centimetres—along almost all of the length of the crystal, and as such diverged much less in this plane. In order to produce a collimated beam a cylindrical high density polyethylene (HDPE) lens of focal length 15 cm was positioned to collimate the quickly diverging vertical axis. Figure 6.9 (left) shows the beam propagation for a 220 μm THz output from the ICTPO after collimation, measured using the scanned knife edge technique. The beam was then focussed through a waist by a 50.8 mm focal length off axis parabolic mirror, shown in Fig. 6.9 (right).

The plots illustrate the high degree of collimation of the THz output that was possible. The beam waist reaches a spot that was only ~2.3 times the diffraction limited spot size. The above data can be used to calculate the M^2 beam quality parameter for the horizontal and vertical cross-sections of the beam to be 6.7 and 2.3, respectively.

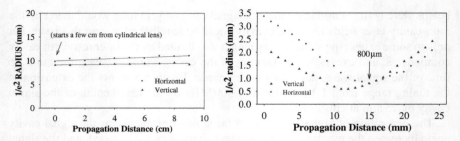

Fig. 6.9 *Left* The beam profile of the THz output collimated by a 15 cm HDPE cylindrical lens. *Right* The propagation of the radiation around the focus of a 50.8 mm focal length lens

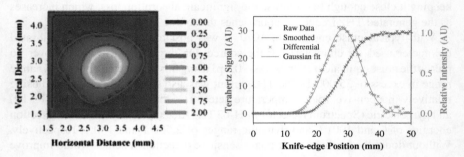

Fig. 6.10 *Left* The beam profile at focus taken using a raster-scanned 100 μm aperture. *Right* An example beam profile taken in a collimated section using the knife edge method

To investigate the beam profile a number of measurements were taken using scanned aperture and knife edge methods, shown in Fig. 6.10, and illustrate the Gaussian, symmetrical nature of the THz output.

These properties are advantageous in imaging systems, and also in spectroscopic systems where the optical path is long (e.g., for stand-off detection systems). An example of the use of the source for imaging is shown in Fig. 6.11 where a piece of salami was rastered through the focus of the beam as the THz power transmitted recorded. Extra detail not visible to the eye is revealed in the fatty sections (most probably areas of high water concentration—a well known strong absorber).

To investigate the quality of the THz beam's polarisation state a set of pulse energy measurements at 1.5 THz were performed while varying the angle of a wire grid polariser that was placed in the beam. Sixty four pulses were averaged on an oscilloscope for each measurement to reduce the standard error to ~ ±2.5 %. The results are plotted in Fig. 6.12 and compared to Malus' law.

The THz wave showed a high degree of polarisation and the wire grid polariser behaved as predicted by Malus' law. The maximum pulse energy transmitted was 6.7 nJ, and the energy at maximum extinction was 4.5 pJ, giving a good extinction ratio of ~1,500:1.

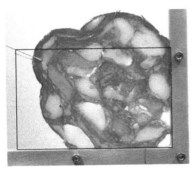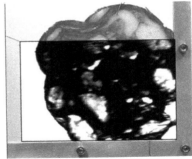

Fig. 6.11 Terahertz transmission at 220 μm (*white* = full transmission, *black* = no transmission) profile (85 by 60 mm) of a sample of salami overlaid with a photograph of the corresponding area in the visible spectrum (backscattered light). The resolution of the raster scan was 250 μm and THz spot size ~0.8 mm radius at e^{-2}

Fig. 6.12 A graph showing the transmission of the metallic grid polariser with varying angle. The *crosses* denote the transmitted pulse energy measured using the bolometer, the *solid line* is a fit to a squared cosine curve (Malus' law) and shows good agreement

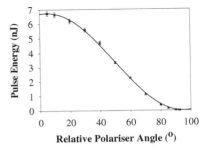

6.9 Line Narrowing

The linewidth of the THz generated by the basic intracavity source described in the previous pages was around 50 GHz, which could be calculated from knowledge of the pump and signal wavelengths and was confirmed via measurements using a Michelson interferometer. This is adequate for a wide variety of spectroscopic measurements of solids and liquids, as these substances generally exhibit significantly broadened spectral lines. There are, however, applications that require narrower linewidths: the spectroscopy of low pressure gasses being an example. The control of the linewidths of the two resonant optical waves by necessity controls the linewidth of the THz wave due to the conservation of energy. As the pump and signal fields are simply optical fields resonant within their respective Fabry-Perot cavities, a host of established line narrowing techniques can be brought to bear in order to realise a narrow linewidth ICTPO. The simplest of these techniques is to place etalons in the resonators of the waves to be narrowed, thereby limiting the range of frequencies supported. This line narrowing method has been demonstrated in the ICTPO [16] and its implementation will now be briefly described.

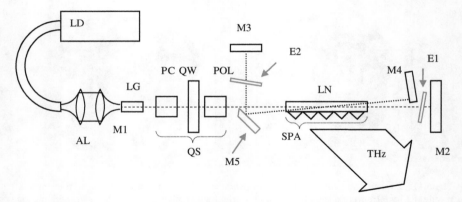

Fig. 6.13 The etalon line narrowed system layout. The system is almost identical to the non-narrowed system outlined in Fig. 6.4 with the modifications highlighted in *red*

The experimental system was a development on the basic system described by Fig. 6.4. All components remained largely unaltered except for the spacing of components which were slightly modified to provide clear spaces for the etalons. The revised system is shown in Fig. 6.13. The etalon used to narrow the laser linewidth (E1) was positioned near the output coupler of the laser. To include an etalon in the signal wave the cavity axis had to be spatially separated further from the pump wave or else the etalon would impinge on both waves. This was accomplished by including a high reflectivity, 90° reflection angle, mirror (M5) where mirror M3 was originally, and moving M3 to retro reflect the signal a few centimetres orthogonally from the laser axis. Angular adjustment of E2 around the vertical axis was realised via a precision galvanometer, allowing the transmission peak of the etalon to be tuned in to match the phase matching solution corresponding to the signal cavity angle, and also enabling the fine tuning of the signal wave within the phase matching bandwidth of the signal cavity.

The etalons were chosen on the basis of their free spectral range (large enough to ensure only one resonant mode) and finesse (too low would not narrow sufficiently, too high results in high round-trip losses). The etalons were chosen from a set of 10, 40 and 80 % surface reflectivities in both 300 and 600 μm thicknesses.

Optimal performance was achieved with etalons E1 and E2 both being 600 μm thick (free spectral range ~166 GHz), and with reflective coatings of 40 and 80 %, respectively (finesse values of approximately 3 and 14, respectively). The effect of etalon E1 on the pulse energy obtained from the free-running Nd:YAG laser (i.e., with parametric down-conversion suppressed) was negligible and easily recoverable by increasing the primary diode pump power. Inserting etalon E2 had a similar effect on the signal wave. As the extracted THz energy was limited by damage to the nonlinear crystal coatings by the resonant signal and, more importantly pump fields, the ICTPO could not be brought as far above threshold, limiting the down conversion efficiency compared to the free running system. For example the system with both etalons reached $1.2 \times$ ICTPO threshold and exhibited $\sim 32\%$ down conversion (measured by monitoring the laser pulse energy with a power meter through the output coupler of the laser cavity), and at a similar primary pumping energy the free running system with only etalon E1 reached 1.4x oscillation threshold and exhibited $\sim 42\%$ down conversion.

The free running linewidth of the laser, in the absence of etalon E1, was measured to be $\sim 20\,$GHz, which narrowed to $<1\,$GHz when etalon E1 was included, as measured by a Fizeau-wedge wavemeter (Angstrom WS-7). Insertion of the etalon E1 in itself effected a reduction in linewidth of the signal wave, bringing it down to $\sim 15\,$GHz on a single pulse basis. When the second etalon (E2) was placed in the signal cavity further reduction of the linewidth down to $<1\,$GHz was obtained. The shot to shot frequency jitter, however, resulted in a broader than expected THz linewidth of $\sim 5\,$GHz. Such a linewidth is compatible with the resolution of individual atmospheric pressure-broadened spectral lines exhibited by gases and vapours [24]. In order to verify the linewidth of the terahertz output as well as demonstrate the utility of the source spectroscopy of the $1.4969\,$THz absorption line of CO gas at reduced pressures was performed. During these measurements the pulse to pulse wavelength of the signal was recorded as the etalon angle was scanned, and was later processed by binning the data at a frequency interval of $250\,$MHz. The pressure of the gas was then reduced and a plot of the measured absorption widths, shown in Fig. 6.14, when traced back to the graph axis indicates a linewidth of $\sim 0.5\,$GHz was achieved with this technique. It is anticipated that optimisation of the selection of signal wave etalon will achieve this linewidth without the need for pulse to pulse wavelength readings.

A summary of the linewidths achievable through the insertion of intracavity etalons into the pump and idler cavities is displayed in Table 6.1.

Considering that the Fourier transform limited minimum bandwidth of a 1 ns pulse is $\sim 100\,$MHz then this etalon technique is clearly very effective and simple to realise. In order to achieve this linewidth in pulsed systems then a different technique is required. In pulsed systems the downconverted waves will build up from random photon noise for every pulse, and so some fluctuation in centre frequency and bandwidth is unavoidable. The solution to this problem is to 'seed' the idler wave of the ICTPO with radiation from a stabilised single frequency laser, forcing the downconverted wave to adopt the spectral properties of this seed wave instead. This technique has been realised in both extracavity and intracavity pumped systems to reach transform limited linewidths of $<100\,$MHz [25]. Although the linewidth achieved was close to transform limited, the system complexity increased significantly over the sub $500\,$MHz etalon-based system, requiring that the laser be operated on single mode via a prelasing technique [26] and that the signal cavity be locked to the seed wave to avoid artificial features to appear in the spectra as the intracavity seed intensity was modulated as shown in Fig. 6.15.

More recent advances in parametric terahertz generation involve the application of quasi phasematching methods in the ICTPO with periodically poled lithium niobate [27]. The goals are to produce simpler devices (where the pump and signal cavities are one and the same), to reach beyond the current tuning range of the TPO which

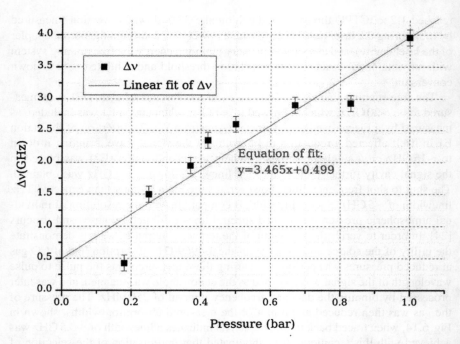

Fig. 6.14 A plot of the linewidth of the 1.4969 THz carbon monoxide rotational transition line as a function of pressure. The *fitted line* indicates a pressure self-broadening coefficient of 3.5 ± 0.5 GHz/bar and an instrumental linewidth of 0.5 GHz

Table 6.1 A summary of averaged FWHM linewidths

Etalons implemented	Pump (GHz)	Signal (GHz)	THz (GHz)
None	20	~50	~50
Pump only	<1	~15	~15
Pump and signal	<1	~5	~5
Pump and signal (single shot)	<0.5	<0.5	~0.5

is currently limited by geometric considerations, and even to improve efficiency by overcoming the Manley–Rowe limit via cascaded downconversion processes. Initial work has highlighted some, non-obvious, difficulties that arise in the case of the quasi phasematched ICTPO due to the inherent bidirectionality of the grating vector component in the phasematching producing dual solutions (see Fig. 6.16).

The bidirectionality is not a difficulty in DFG as both waves are present from the start, but in the case of the ICTPO the downconverted waves build up from noise and can take either solution. Higher gain is anticipated for the solution with the least walk off of the waves, and unless carefully designed this results in unextractable THz radiation. Methods to produce successful devices have been put forward in [27].

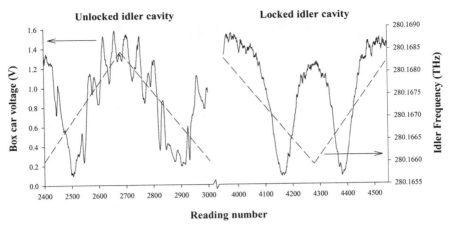

Fig. 6.15 A graph illustrating the effect of locking the idler cavity to the seed frequency while tuning over the 1.4969 THz absorption line at 80 mbar. The *dashed line* shows the seed laser frequency as measured by wavemeter and the *solid line* the THz energy recorded by composite silicon bolometer. Note there are also discontinuities caused by mode hops of the ICTPO pump laser in the unlocked data

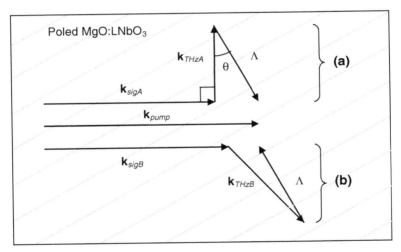

Fig. 6.16 A k-vector diagram illustrating the dual solutions that exist in the quasi phase matching ICTPO due to the bidirectionality of the grating vector (Λ). Here the quasi phasematching solution was originally designed for orthogonal exit of the THz wave (**a**), however, a second solution with a different terahertz frequency and propagation direction is also possible (**b**)

6.10 Application of Intra-Cavity THz OPO in Stand-off Spectroscopy

6.11 Introduction

This chapter has so far introduced a number of approaches for nonlinear generation of THz radiation. This section will give a brief overview on how THz can be utilised in the field of standoff spectroscopy. This area is currently dominated by a number of techniques based on ultra violet (UV), visible and infrared (IR) laser sources; however, these spectroscopy techniques are limited as they do not penetrate clothing/materials [28, 29]. THz radiation can overcome this limitation as it can penetrate barriers like clothing, paper and plastic with moderate attenuation [30–33]. Indeed one application where THz spectroscopy has been proposed is mail screening for explosives and drugs [34], as packages can be investigated without the need for them to be opened. Compared to X-rays that also can penetrate these barriers, THz radiation has a relatively low photon energy (4 meV at 1 THz) that is safer for living tissues and DNA. Moreover, many compounds of interest, such as energetic materials, have spectroscopic features in the THz band [35–42]; THz offers a powerful alternative for stand-off detection in a variety of application areas. THz radiation has the ability to probe long range intermolecular forces between molecules within a crystal lattice, which allows spectra of materials to be produced. In comparison for the IR region it is the intramolecular vibrations between the atoms in the molecules that produce the spectra.

This overview of stand-off spectroscopy with focus on explosives detection, as a wide range of work has already being carried out and detection techniques are currently in demand, especial for improvised explosive devices (IEDs) and for security check points in controlled areas, such as at airports, sport stadiums or justice courts. A number of groups have published THz spectra for a range of common energetic compounds (RDX, PETN, HMX and TNT) and commercial explosive compounds (PE4 and Semtex-H), showing that there are a number of unique features in the 0–4 THz range [35–42], illustrated by Fig. 6.17.

The potential of the THz spectroscopy to identify energetic substances also relies on the uniqueness of the spectral features compared with compounds of *non-interest*, such as sugar, flour or chocolate. The discrimination between energetic substances with compounds that are not of interest is required to avoid a high false positive rate, meaning that too many substances would be identified as dangerous. Figure. 6.18 shows the THz spectra of a number of foodstuffs [33]. Some of these foodstuffs could be used as confusing material. To aid in the successful identification of compounds, either the identification software tool must be precise and/or the THz source should have a wide frequency range, which will allow more spectral features to be observed.

To achieve useable spectroscopic measurements the system must have a high SNR to overcome signal attenuation due to losses mentioned previously. A portion of these losses will be determined by the nature of the scattering/reflections, for example diffuse reflections have been found to be 20–30 dB weaker than specular

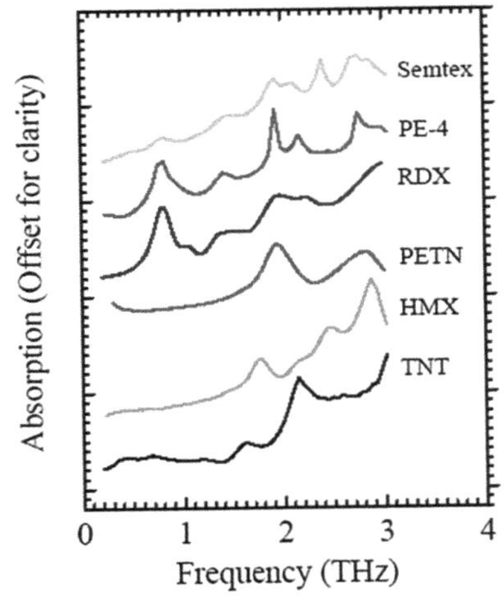

Fig. 6.17 Typical THz spectra for a selection of energetic and explosive compounds [39]

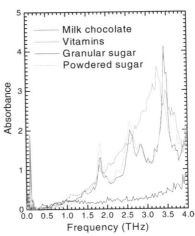

Fig. 6.18 Typical THz spectra for a selection of food stuffs (compounds of non-interest) [33]

reflections [43]. Further discussions on the effect of scattering on THz reflection spectra can be found at [44–48].

6.12 Potential of the ICTPO for Standoff THz Spectroscopy

The potential for identification of energetic material hidden in a container or under clothing was quickly recognized after the breakthrough [49, 50] in emission and detection of THz radiation that allowed affordable table top systems. Although many published results demonstrate the potential of THz for concealed threat detection, the development of a system that can operate in a standoff manner, with hidden explosive and in a real environment (outside the laboratory), is a challenge [51]. First, the THz radiation is absorbed by water vapour [52] present in the atmosphere, especially above 0.5 THz. Although the water absorption is limited to known lines in the THz spectrum, propagating over long distance requires having high peak power THz source and/or limiting the system operation to transmission windows only. The ICTPO is a suitable source for long distance THz propagation as it can generate high peak power, narrow band pulses that can be tuned within the transmission windows of the atmosphere.

Second, the parts of the spectrum that are considered transparent show, in reality, residual absorption and/or moderate scattering at THz frequencies, but not uniformly across the spectrum. This adds features to the detected spectrum that complicates the identification of the target substance. A number of THz TDS-based studies have measured the absorption spectra for several common *barrier* materials [30–33], however, there remain discrepancies in the results. It was demonstrated that these discrepancies are due to the non-uniformity in the structure of the barrier sample used that cause spatial coherence distortions in the detection process used for THz TDS [53]. An example of different spectral absorption results for polyester is shown in Fig. 6.19. The measurement shows different results for the polyester under the form of a structural mesh or in a powder. This problem is observable when measurements are performed with a THz TDS system. Such systems are based on wide band THz pulses and by necessity use coherent detection techniques. More details on THz TDS are presented in Chap. 8.

THz spectrometers based on an ICTPO as the source and a thermal detector such as a Golay cell or a bolometer, do not rely on coherence. Such a spectrometer does not experience the spatial coherence distortion present with TDS systems. The drawback of using nanosecond THz pulses generated by the ICTPO is that it is not possible to separate the reflection of the substance of interest from that of the barrier, if present, due to the insufficient temporal resolution, complicating the identification of the substance. With THz TDS, under certain conditions, time gating can remove

Fig. 6.19 The influence of the barrier material on THz spectral showing the graph (**a**) for polyester (**b**) and grinded polyester (**c**)

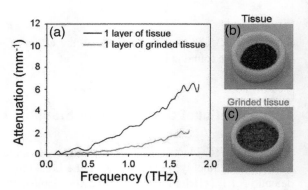

the spectral features of the reflection of the barrier, although care must be taken to ensure this does not effect spectral content and resolution.

Other difficulties for substance identification by THz spectrometry are the surface roughness, surface curvature and orientation of the sample. They attenuate the specular reflection and add to the scattering of the THz radiation on the detector, degrading spectrum [44, 46–48, 54]. Again, the peak power of the THz source is capital to overcome this challenge.

6.13 THz Spectroscopy Based on OPO Source

The potential of THz spectroscopy for standoff detection and identification of energetic substances still needs to be evaluated in real operational conditions. This issue is being addressed by a number of research groups. The Defence Research and Development Canada (DRDC) is part of an international group that is addressing this issue by performing experimental investigation. The DRDC developed a new THz spectroscopy system to demonstrate the identification of hidden energetic material at a standoff distance of more than a metre in normal relative humidity conditions.

This unique THz spectroscopy system used an intracavity OPO THz source with a linewidth of <50 GHz and a THz pulse energy of >30 nJ, giving a peak power of >3 W in the THz region. A simple telescope formed of two lenses was used to focus the THz beam at the desired distance; one of the lenses was mounted on a computer-controlled translation stage that could vary the focal plane of the telescope. A folding mirror was used to aim the THz beam on target. To ease the THz beam pointing, a visible laser beam boresighted with the THz beam was imlemented. A removable mirror was used before the telescope to couple the visible beam in the optical path of the THz. The detection was performed with a bolometer cooled down to 4.2 K using liquid helium. This kind of detector is not suitable for field use, but subsequent developments in the technology now allow bolometers to be produced with the same performance that do not require liquid helium cooling. The bolometer was mounted on a two-axis computer controlled translation stage to optimise the detection of the reflected THz on the target. The control of the system (source emission frequency, alignment of the bolometer, adjustment of the beam focus) and the data collection were performed on a computer via custom software. A schematic diagram of the system is shown in Fig. 6.20.

The first prototype of this THz spectrometer was used in an international experiment with the objective of evaluating the THz systems outside the laboratory controlled conditions [55]. The OPO based THz system demonstrated the identification of substance positioned at 8 m, meaning that the THz radiation had to propagate through a round trip path of 16 m. A typical result is shown in Fig. 6.21 . It shows the THz spectrum reflected from a sample of tartaric acid (an explosive simulant [56]). The characteristic drop in the spectrum around 1.2 THz of the tartaric acid is clearly observable. When the sample is hidden behind layers of polyester it can still be seen.

Fig. 6.20 A schematic diagram of the THz standoff spectrometer system using an OPO THz source

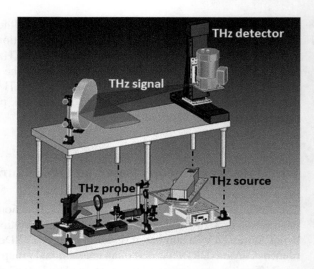

Fig. 6.21 The spectra recorded of tartaric acid at a standoff distance of 8 m with and without barriers of polyester

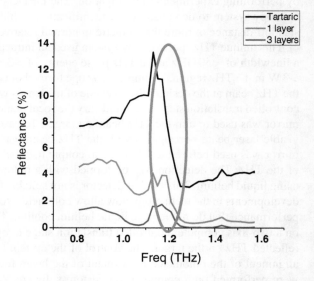

A combination of increased pulse energy and spectral brightness permitted an increase of one to two orders of magnitude in the standoff distance for identifying energetic material hidden behind barriers compared to other sources/detection techniques. A source of this nature is suitable to compose with the barrier of atmospheric absorption for standoff THz spectroscopy and imaging. It does not overcome all the challenges of detecting and identifying substances in real applications, as for example the discrimination of a reflection from a barrier, but it gives a roadmap of how the field of THz spectroscopy can mature and produce a product that take advantage of the main benefits offered by the THz technology; penetration of barriers and safety for the living tissues.

6.14 Conclusions

The intracavity approach has led to a significant reduction in power requirements for the ICTPO, making them particularly practical sources of THz radiation, and has enabled the development of a compact and robust commercial THz source that will aid researchers in academia and industry exploit this little used frequency range. The generated THz radiation has been shown to exhibit a very good Gaussian-like beam profile, low divergence and a high degree of polarisation, as well as a high spectral brightness and a broad and continuous tuning range in the middle of the so-called THz gap. This combination of features is not existent in any other practical THz source, giving the ICTPO the potential to be instrumental in developing THz science and applications.

An example of this potential was given with a brief overview of THz stand-off spectroscopy with reference to security applications. The overview has highlighted how a THz source can be used to penetrate clothing and other materials in order to probe a concealed compound such that it may be identified. A number of factors that can affect performance of a standoff system have been identified including: atmospheric absorbance, barrier attenuation and reflection/scattering losses. For the practical implementation of a standoff system the above factors must be considered, along with the operating specification of the THz laser source, such that meaningful and useable results are obtained. The properties of the THz radiation (spectral brightness, tunability, bandwidth, and peak power) from an ICTPO are shown to be advantageous for this application.

Acknowledgments Thanks go to Dr D.J.M. Stothard for his work in collecting the data and producing the figures for the THz beam profiling and imaging demonstration and to Dr T.J. Edwards and C.L.Thomson for providing figures for the temporal traces of the ICTPO pulses and the variation of absorption width of the 1.4969 THz absorption line respectively. Finally, M.H. Dunn is thanked for his assistance in producing this work.

References

1. G.K. Kitaeva, Laser Phys. Lett. **5**, 559 (2008)
2. D. Creeden et al., Opt. Express **15**, 6478 (2007)
3. Y. Jiang, Y.J. Ding, Appl. Phys. Lett. **91**, 091108 (2007)
4. T. Taniuchi, H. Nakanishi, J. Appl. Phys. **95**, 7588 (2004)
5. K.L. Vodopyanov, Laser Photonics Rev. **2**, 11 (2008)
6. J. Gelbwachs et al., Appl. Phys. Lett. **14**, 258 (1969)
7. J.M. Yarborough et al., Appl. Phys. Lett. **15**, 102 (1969)
8. M.A. Piestrup, R.N. Fleming, R.H. Pantell, Appl. Phys. Lett. **26**, 418 (1975)
9. K. Kawase et al., Appl. Phys. Lett. **68**, 2483 (1996)
10. K. Kawase et al., Appl. Opt. **40**, 1423 (2001)
11. K. Kawase et al., Appl. Phys. Lett. **80**, 195 (2002)
12. L. Palfalvi et al., J. Appl. Phys. **97**, 123505 (2005)
13. R.W.P. Drever et al., Appl. Phys. B: Lasers Opt. **31**, 97 (1983)

14. M. Ebrahimzadeh, M.H. Dunn, in *Handbook of Optics IV: Fiber and Nonlinear Optics*, ed. by M. Bass, J.M. Enoch, vol. **4** (Optical Society of America, 1998), p. 22.1
15. T. Edwards et al., Opt. Express **14**, 1582 (2006)
16. D.J.M. Stothard et al., Appl. Phys. Lett. **92**, 141105 (2008)
17. K. Kawase, J.-I. Shikata, H. Ito, J. Phys. D: Appl. Phys. **35**, R1 (2002)
18. D.A. Bryan, G. Robert, H.E. Tomaschke, Appl. Phys. Lett. **44**, 847 (1984)
19. J. Shikata et al., IEEE Trans. Microw. Theory Tech. **48**, 653 (2000)
20. D.E. Zelmon, D.L. Small, D. Jundt, J. Opt. Soc. Am. B **14**, 3319 (1997)
21. T.Y. Fan, R.L. Byer, IEEE J. Quantum Electron. **24**, 895 (1988)
22. M.H. Dunn et al., Internal memos on the analysis of gain in the (Intracavity) Terahertz Parametric Oscillator
23. S. Brosnan, R. Byer, IEEE J. Quantum Electron. **15**, 415 (1979)
24. L.S. Rothman et al., J. Quant. Spectrosc. Radiat. Transf. **110**, 533 (2009)
25. D. Walsh et al., J. Opt. Soc. Am. B **26**, 1196 (2009)
26. D.C. Hanna et al., Opt. Quantum Electron. **3**, 163 (1971)
27. D.A. Walsh et al., Opt. Express **18**, 13951 (2010)
28. S. Wallin et al., Anal. Bioanal. Chem. **395**, 259 (2009)
29. J. Yinon (ed.), *Forensic and Environmental Detection of Explosives* (Wiley, Chichester, 1999)
30. J.E. Bjarnason et al., Appl. Phys. Lett. **85**, 519 (2004)
31. A.J. Gatesman et al., in *Terahertz for Military and Security Applications IV*, ed. by D.L. Woolard, et al. (SPIE, Orlando (Kissimmee), FL, 2006), p. 62120E
32. M.C. Kemp, C. Baker, I. Gregory, in *Stand-off Detection of Suicide Bombers and Mobile Subjects*, ed. by H. Schubert, A. Rimski-Korsakov (Springer, New York, 2006), p. 151
33. W.R. Tribe et al., in *Integrated Optoelectronic Devices 2004*, (International Society for Optics and Photonics, 2004), p. 168
34. H. Hoshina et al., Appl. Spectrosc. **63**, 81 (2009)
35. A. Burnett et al., in *Optics and Photonics for Counterterrorism and Crime Fighting II*, ed. by C. Lewis, G.P. Owen (SPIE, 2006), p. 64020
36. Y. Chen, H. Liu, X.-C. Zhang, in *Terahertz for Military and Security Applications IV*, ed. by D.L. Woolard et al. (SPIE, 2006), p. 62120P
37. D.J. Cook et al., in *Terahertz and Gigahertz Electronics and Photonics III*, ed. by R.J. Hwu (SPIE, 2004), p. 55
38. F. Huang et al., Appl. Phys. Lett. **85**, 5535 (2004)
39. M.C. Kemp et al., in *Terahertz for Military and Security Applications*, ed. by R.J. Hwu, D.L. Woolard (SPIE, 2003), p. 44
40. M.R. Leahy-Hoppa et al., Chem. Phys. Lett. **434**, 227 (2007)
41. Y.C. Shen et al., Appl. Phys. Lett. **86**, 241116 (2005)
42. K. Yamamoto et al., Japan. J. Appl. Phys. **43**, L414 (2004)
43. H.B. Liu et al., Opt. Express **14**, 415 (2006)
44. M. Herrmann et al., in *Terahertz Physics, Devices, and Systems III: Advanced Applications in Industry and Defense*, ed. by M. Anwar, N.K. Dhar, T.W. Crowe (SPIE, 2009), p. 731105
45. H. Richter et al., in *Terahertz Physics, Devices, and Systems III: Advanced Applications in Industry and Defense*, ed. by M. Anwar, N.K. Dhar, T.W. Crowe (SPIE, 2009), p. 731104
46. D.M. Sheen, D.L. McMakin, T.E. Hall (eds.), in *Defense and Security Symposium* (International Society for Optics and Photonics, Orlando, 2007), p. 654809
47. J. Wilkinson, S.M. Caulder, A. Portieri, in *Terahertz for Military and Security Applications VI*, ed. by J. O. Jensen et al. (SPIE, 2008), p. 694904
48. L.M. Zurk et al., in *Terahertz for Military and Security Applications VI*, ed. by J. O. Jensen et al. (SPIE, 2008), p. 694907
49. D.H. Auston, Appl. Phys. Lett. **26**, 101 (1975)
50. R.A. Cheville, D. Grischkowsky, Appl. Phys. Lett. **67**, 1960 (1995)
51. A.D. Van Rheenen, M.W. Haakestad, in *Detection and Sensing of Mines, Explosive Objects, and Obscured Targets XVI*, ed. by R.S. Harmon, J. John H. Holloway, J.T. Broach (SPIE, 2011), p. 801719

52. R. Appleby, H.B. Wallace, IEEE Trans. Antennas Propag. **55**, 2944 (2007)
53. F. Theberge et al., Opt. Express **17**, 10841 (2009)
54. M. Ortolani et al., in *Terahertz Physics, Devices, and Systems III: Advanced Applications in Industry and Defense* (SPIE, 2009), p. 7329
55. Multiple, NATO Research and Technology Organisation RTO-TR-SET-124 AC/323(SET-124)TP/399 (2011)
56. C. Konek et al., in *Terahertz Physics, Devices, and Systems III: Advanced Applications in Industry and Defense*, ed. by M. Anwar, N.K. Dhar, T.W. Crowe (SPIE, 2009), p. 73110K

Part III
Systems and Applications

Chapter 7
Terahertz Control

David R. S. Cumming, Timothy D. Drysdale and James P. Grant

Abstract Light control is essential to the design and implementation of optical systems. Basic properties that can be manipulated include the spectrum, polarisation and focusing. In this chapter we will look at a range of technologies for the manipulation of terahertz radiation, their properties and their method of implementation.

Keywords Metamaterials · Diffractive optics · Terahertz optics · Terahertz filters · Tunable filters · Artificial dielectrics · Variable artificial dielectric retarder · Photonic bandgap filter

Light control is essential to the design and implementation of optical systems. Basic properties that can be manipulated include the spectrum, polarisation and focusing. In this chapter we will look at a range of technologies for the manipulation of terahertz radiation, their properties and their method of implementation.

7.1 Diffractive Optics

The traditional lens is most commonly based on a refractive element that exploits the change in the phase velocity of a wave as it moves from a medium of one refractive index into another. The most basic property of a lens—its focal length—is determined by the radius of curvature of the faces of the lens. For a lens with a short focal length the radius of curvature must be relatively small, making it a so-called thick lens. As a consequence, the size and weight of a simple single-element lens is determined by

D. R. S. Cumming (✉) · T. D. Drysdale · J. P. Grant
School of Engineering, University of Glasgow, Rankine Building, Oakfield Avenue, Glasgow
G12 8LT , UK
e-mail: d.cumming@elec.gla.ac.uk

M. Perenzoni and D. J. Paul (eds.), *Physics and Applications of Terahertz Radiation*,
Springer Series in Optical Sciences 173, DOI: 10.1007/978-94-007-3837-9_7,
© Springer Science+Business Media Dordrecht 2014

Fig. 7.1 Refraction at a
spherical surface

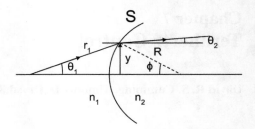

its aperture (the lens diameter) and focal length. Large aperture lenses are therefore
heavy, especially in applications where a large numerical aperture (NA) is required.

Instead of making a refractive lens, a diffractive element known as a zone-plate
may be made instead. Diffractive lenses may be optically thick whilst simultaneously
being physically thin, hence light in weight. Diffractive lenses also lend themselves
well to low cost manufacturing methods. The cost for using a diffractive optic is that
it is inherently dispersive so that the focal length is dependent on the radiation wavelength. Consequently, imaging systems, for example, are typically monochromatic.

7.1.1 Design of a Lens

A ray of light on trajectory (y, θ_1) at a distance y from the optical axis of a spherical boundary, S, passes from a medium of refractive index n_1 into a medium with
refractive index n_2. The ray will be refracted on to a new trajectory (y, θ_2) as shown
in Fig. 7.1. The spherical boundary has a radius R.

The relationship between the two ray trajectories is given by a transfer matrix so
that

$$\begin{bmatrix} \theta_2 \\ y \end{bmatrix} = \begin{bmatrix} n_1/n_2 & \frac{n_1-n_2}{n_2 R} \\ 0 & 1 \end{bmatrix} \begin{bmatrix} \theta_1 \\ y \end{bmatrix}. \tag{7.1}$$

This essential construct can be developed to show that for a lens made of a material
with a refractive index n and two surfaces with radii of curvature R_1 and R_2, the
focal length of the lens will be

$$f = \frac{1}{n-1} \left(\frac{1}{R_1} - \frac{1}{R_2} + \frac{(n-1)l}{n R_1 R_2} \right)^{-1} \tag{7.2}$$

where l is the thickness of the lens at its centre [1]. For a simple bi-convex lens with
$R_1 = R_2$, this can be further reduced to

$$f = \frac{n R_1^2}{l(n-1)^2} \tag{7.3}$$

Fig. 7.2 a The projection of rings of constant phase at a spherical surface on to a plane. **b** An amplitude modifying pattern mapped on to the zone boundaries and **c** a phase modifying design

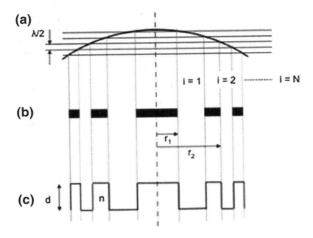

Alternatively, we may choose to design a plano-convex lens that will be simpler to convert into a zone-plate. In such a case, one of the surface radii (e.g. R_2) will tend to infinity so that

$$f = \frac{R_1}{n-1}. \tag{7.4}$$

Fresnel zones are rings observed in a diffraction pattern from a pin-hole. The observed rings have well-defined radii. By reciprocity, a zone-plate is a circular grating-like structure with grooves positioned at radii determined by the Fresnel zones. For a spherical plano-convex lens the zone radii can be determined by mapping disks of constant phase on to a plane. This is shown in cross-section in Fig. 7.2 where the wavefronts are separated by half the wavelength.

The resulting zone radii can be shown to be

$$r_i = \sqrt{if\lambda + \frac{i^2\lambda^2}{4}} \tag{7.5}$$

where i is the zone number, f is the focal length, and λ is the wavelength. The simplest way to make a zone-plate is to fill out half of each zone with an absorber (e.g. metal on glass) to make an amplitude modifying grating (Fig. 7.2b). However, this method will block half the light reducing the total transmission to less than 50 %. The light focussed into the first order diffraction mode that determines the focal length will be only 22 %. It is therefore better to make a phase modifying zone-plate (Fig. 7.2c). For such a lens, the whole area is transmissive, and for a simple binary design with two phase levels separated by one half wavelength, a grating must be etched or cut into a suitable dielectric medium to a depth d that is given by

Fig. 7.3 a An ideal spherical profile when added to a phase modifying zone-plate will make a lens that transmits all the light into the first-order mode, and eliminate all negative modes. **b** A spherical profile can be approximated by a stepped profile that is easier to make. In the example shown there are four phase levels

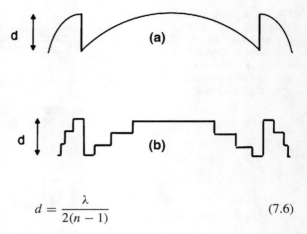

$$d = \frac{\lambda}{2(n-1)} \tag{7.6}$$

where n is the refractive index of the material into which the lens is made. A lens made in this way will have an efficiency of 44%. The rest of the light will be scattered into higher order modes. To make a more efficient lens the high order modes must be suppressed, and this can be done by introducing a blaze. The effect of the blaze is to break the symmetry of the diffraction pattern so that all negative modes are eliminated, and only the first-order positive mode is propagated. Ideally, a blaze would follow a spherical contour as shown in Fig. 7.3a. In practice, even for the relatively large geometries entailed in working at terahertz wavelengths, perfect spherical contouring is difficult to achieve, hence a multilevel binary blaze is more suitable. This is illustrated in Fig. 7.3b where a stepped structure with 2^m phase levels is used. Such a structure can be patterned in only m successive mask and etch steps, making it especially appropriate to the use of MEMS-based techniques.

7.1.2 Examples of Terahertz Diffractive Optics

Silicon micromachined diffractive optics have been demonstrated as shown in Fig. 7.4. The lens shown is designed to have a focal length of 25 mm at 1 THz. The lens is fabricated using successive photolithography and etch processes. The etch was made using an inductively coupled plasma deep-reactive ion etching process. The same basic lens design was made with different numbers of phase steps, and the results of Fig. 7.5 show clearly how greater efficiency can be achieved with more phase levels [2–7]. Because diffractive optics are dispersive, the focal length will vary as a function of the wavelength. Using this property it has been possible to demonstrate a dispersive tomographic imaging system [8]. Holograms have also been demonstrated since they permit arbitrary light patterns to be formed using a single optical element [9–11].

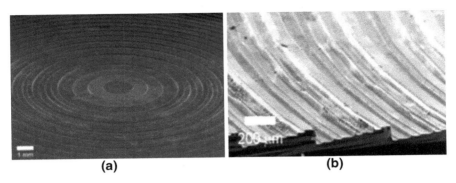

Fig. 7.4 a The surface of a terahertz Fresnel zone-plate and **b** a cross-section showing an eight-level blaze structure

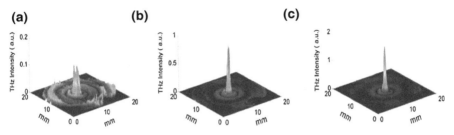

Fig. 7.5 The beam intensity profiles at the focus of three terahertz lenses at 1 THz illustrating the efficiency that is obtained from **a** 2-levels, **b** 4-levels, and **c** 8-levels of phase. The 8-level lens is c. 92 % efficient [9]

7.2 Polarisation Control

Polarisation is a property of light that is frequently manipulated in imaging, communication and system design. A commonplace polarisation altering device is the polariser. This is a simple device in that it selects only light on a single linear polarisation so that light after the polariser is always linearly polarised. A more sophisticated device is the waveplate which can retard light of orthogonal polarisation vectors by a different amount. The most common example is the quarter-waveplate that can convert linearly polarised light to circularly polarised light (and back again). Using this property it is possible to make an isolator—an essential component of many systems.

7.2.1 Basic Theory

A propagating beam of light may be described by the following equation:

$$\vec{E}_x = A_x e^{j(kz-\omega t)}. \tag{7.7}$$

Equation (7.7) describes a wave with its electric field aligned to the x-axis of a Cartesian space that is propagating in the z-direction with a wavevector $k = 2\pi/\lambda$, at a frequency ω, or wavelength λ. The wave is linearly polarised. More generally we may write:

$$\begin{aligned} \vec{E} &= (\vec{E}_x, \vec{E}_y, \vec{E}_z) \\ &= (A_x e^{j(kz-\omega t)}, A_y e^{j(kz-\omega t+\phi)}, 0) \\ &= (A_x, A_y e^{j\phi}, 0) e^{j(kz-\omega t)} \end{aligned} \tag{7.8}$$

This is a wave with the electric component resolved into three components E_x, E_y, and E_z. $E_z = 0$ since there is no electric field in the direction of propagation. However, both E_x and E_y may be non-zero, with independent amplitudes A_x and A_y. The electric field on each of the x and y-axes may be in phase, or separate in phase by the polarisation angle, ϕ. When E_x and E_y are perfectly in phase the light is linearly polarised, but if $\phi = 90°$, then the phase front rotates as the wave propagates and the light is said to be circularly polarised if $A_x = A_y$.

The angle ϕ, hence the polarisation, can be altered by light scattering, and reflection from a plane mirror can introduce well controlled rotation. However, a device such as a waveplate is less cumbersome and more commonplace. For a waveplate to work it must be birefringent—that is to say, the field components on different axes (e.g. E_x and E_y) must experience a different refractive index, and hence one will be retarded with respect to the other. This can be achieved using a birefringent crystal such as calcite [12]. Equation (7.8) becomes

$$\vec{E} = (A_x, A_y e^{j\frac{2\pi(n_y-n_x)l}{\lambda}}, 0) e^{j(kz-\omega t)} \tag{7.9}$$

To make a quarter-waveplate the thickness of the calcite, with a given n_x and n_y, must be carefully chosen to make $\phi = 90°$ exactly.

7.2.2 The Babinet Compensator

In many applications a degree of tunability is required. A traditional waveplate is not tunable and may offer only very narrowband operation. To overcome this a variable compensator may be used [13]. Fig. 7.6 shows a Babinet compensator that is made from two Wollaston prisms that are cut with the crystal orientation at right angles with respect to one another. One prism is placed on top of the other so that they can slide over each other. Depending on the relative position of the two prisms the optical path length is varied. If one prism is fixed in position with respect to the propagating beam of light, then the retardation of the polarisation can be controlled very precisely as the second prism is moved.

Fig. 7.6 A sketch of a Babinet compensator made using two birefringent prisms with crystal axes set at 90° to one another

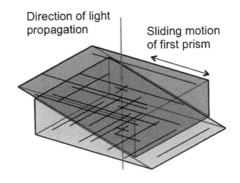

Direction of light propagation

Sliding motion of first prism

The same concepts can be applied to waves of any frequency, but at longer wavelengths the material thickness for the waveplate will be larger. For very high frequency electronics (e.g. micro and millimetre wave) it can also become difficult to find appropriate high-quality single crystals of sufficient size and transparency [12], so alternative solutions such as artificial dielectric methods become attractive.

7.2.3 Artificial Dielectrics

Artificial dielectrics are made by cutting grooves into the surface of a suitable dielectric that is not birefringent, and preferably non-dispersive across the wavelength range of interest. If the surface pattern is a regular two-dimensional grid the structure can be used to make an antireflection layer using the so-called moth-eye effect [14], but if the pattern is a one-dimensional linear grating, then it will have the property of birefringence [15, 16]. Consider the cross-section of a grating shown in Fig. 7.7a. The grating has a depth h, a period Λ and a fill-factor given by the ratio $F = a/\Lambda$. If the electric field is polarised perpendicular (transverse electric or TE mode) or parallel (transverse magnetic or TM mode) to the grating vector K then the effective dielectric constant experienced by a wave propagating through the media will be

$$\varepsilon_{TE} = \varepsilon_1 F + \varepsilon_2 (1 - F)$$

(7.10)

$$\varepsilon_{TM} = \left(\frac{F}{\varepsilon_1} + \frac{(1 - F)}{\varepsilon_2} \right)^{-1}$$

where ε_1 and ε_2 are the dielectric constants of the media as illustrated in Fig. 7.7a. The dielectric functions as a function of the fill-factor are illustrated in Fig. 7.7b.

The difference in the effective dielectric constants for the TE and TM modes may be used in exactly the same way as a bulk birefringent material would be (Eq. 7.10), but with the advantage that cheap and readily available dielectrics can be used. Fig. 7.8 shows a rectangular grating structure made in silicon that can be used as a waveplate

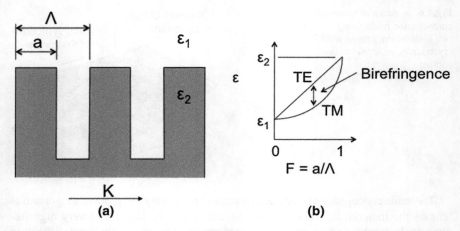

Fig. 7.7 **a** A cross-section through a linear grating made into a dielectric and **b** sketches of the effective dielectric constants for TE and TM modes as a function of the fill-factor a/Λ

Fig. 7.8 A scanning electron micrograph of a rectangular grating in silicon to make a terahertz artificial dielectric

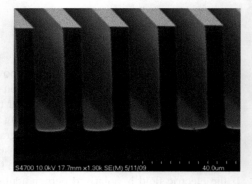

at terahertz wavelengths. The device is made by photolithography and deep-reactive ion etching [17].

7.2.4 Examples of Polarisation Controllers at Terahertz

Artificial dielectrics do not lend themselves well to making Babinet compensators, but there is an alternative micromechanical structure that can exploit the birefringence of a grating. Figure 7.9a shows how two gratings, this time with a triangular surface profile, can be interlocked [18]. When the two gratings are perfectly mated and there is no gap between them, then there is no artificial dielectric, hence no birefringence. However, when the two gratings are separated a birefringence that varies as a function of the separation is created (Fig. 7.9b). Whilst there is an overlapping interlock

Three VADR devices with increasing separation to the right

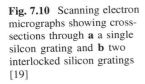

(a)   **(b)**   **(c)**

No Separation S = 0   Partial separation S = S$_B$ Full Separation S = S$_c$

0°Retardance   45°Retardance   90°Retardance

Fig. 7.9 A variable artificial dielectric retarder (VADR) for use at terahertz frequencies. **a** Two plates fully interlocked, **b** partial separation and **c** full separation

Fig. 7.10 Scanning electron micrographs showing cross-sections through **a** a single silcon grating and **b** two interlocked silicon gratings [19]

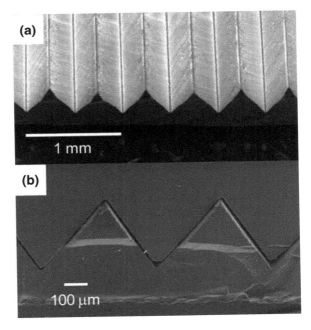

between the gratings the birefringence, hence retardation of the polarisation, grows with the separation. For larger separation the device becomes a Fabry-Perot cavity with birefringent facets (Fig. 7.9c), and the retardation varies cyclically as a function of the separation. An electron micrograph of two silicon gratings designed for use at 100 GHz is shown in Fig. 7.10 [19]. The simulated and experimentally determined behaviour of the device is shown in Fig. 7.11.

The triangular structures are not only desirable because they allow for an easily interlocking structure, but because they lend themselves very well to fabrication in

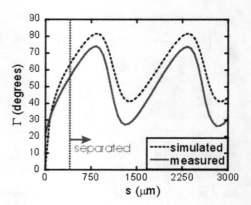

Fig. 7.11 The simulated and experimentally obtained characteristics of the VADR device as a function of the separation between the silicon plates

silicon. The device shown in Fig. 7.10 was made by patterning a linear grating into a silicon nitride etch mask on the surface of a <100> silicon wafer. Wet etching with KOH naturally gives rise to the triangular profile since the etch stops on the <111> plane. Improved performance is possible by modifying the device's cross sectional profile [20].

7.3 Terahertz Filters

The term "filter" is usually taken to mean an optical or electronic device that selects for wavelength or frequency. There are numerous methods for implementing filters [21–24], and one that has captured a lot of interest in optics and at terahertz frequencies is the photonic or electromagnetic bandgap (PBG or EBG, respectively) filter, for example [25–31]. As with the variable artificial dielectric retarder, EBG structure lend themselves well to micromechanical fabrication methods, and can be made to be tuneable.

7.3.1 Elementary Theory of EBG

To understand EBG we must first appreciate the effect of a regular crystal wave that is propagating, or attempting to propagate, within it. When a forward travelling wave enters a crystal it will meet a succession of dielectric interfaces between the materials that make up the structure. The wave will experience some reflection at each interface so that there will also be a backward travelling wave in the crystal. The full wave solution is then the sum of the forward and backward travelling waves. This is illustrated in Fig. 7.12a.

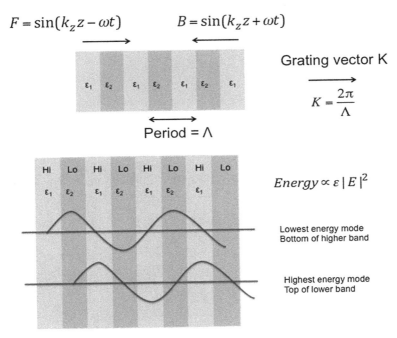

$$F = \sin(k_z z - \omega t) \qquad\qquad B = \sin(k_z z + \omega t)$$

Grating vector K

$$K = \frac{2\pi}{\Lambda}$$

Period = Λ

$$Energy \propto \varepsilon\,|\,E\,|^2$$

Lowest energy mode
Bottom of higher band

Highest energy mode
Top of lower band

Fig. 7.12 **a** A 1D crystal with forward (F) and backward (B) travelling waves, and **b** two extreme possibilities for standing waves in the crystal

The addition of the forward and backward waves gives rise to a standing wave, as shown in Eq. (7.11).

$$
\begin{aligned}
E_y = F + B &= \sin(k_z z - \omega t) + \sin(k_z z + \omega t) \\
&= 2\cos(\omega t)\sin(k_z z)
\end{aligned}
\tag{7.11}
$$

Figure 7.12b illustrates how the standing waves might appear if they are phase matched to the crystal. When they are well matched, there will be no propagating mode: this is the origin of the bandgap. Fig. 7.12b shows two possible conditions. When the intensity of the electric field has its maxima in the material with the lowest dielectric constant, then the energy stored in the standing wave will be at its least. When the intensity of the electric field has its maxima in the material with the highest dielectric constant then energy stored in the material will be at its highest. These two conditions set the upper and lower band edges of the bandgap, respectively.

To calculate the band diagram, more normally thought of as the dispersion relation in optics, we must solve the wave equation. This is the so-called curl-curl equation, written for the electric field in Eq. (7.12).

$$\frac{1}{\varepsilon_r} \nabla \times \nabla \times \vec{E} = -\frac{1}{c^2} \frac{\partial^2 \vec{E}}{\partial t^2} \tag{7.12a}$$

$$\frac{1}{\varepsilon_r} \nabla \times \nabla \times \vec{E} = \left(\frac{\omega}{c}\right)^2 \vec{E} \tag{7.12b}$$

where E is the electric field, ε_r is the real-space distribution of the crystal dielectric, c is the speed of light in vacuo, and t is time. Equation (7.12b) is the frequency-domain version of the equation that explicitly shows the eigenfunction E and its eigenvalues, ω. We are interested in the modes of the eigenvalues that determine the frequency response of an EBG filter.

For a crystal-like dielectric structure that repeats periodically in real-space, r, we may deploy Bloch's theorem, thus

$$\varepsilon(x + nr) = e^{jnk.r} \varepsilon(r). \tag{7.13}$$

where n is an integer. The implication is that the crystal is the same everywhere (i.e. periodic) in each unit cell. The electric field, therefore, in k-space has solutions such that

$$\vec{E}(k) = \vec{E}(k + nK) \tag{7.14}$$

where K is the crystal vector.

Unsurprisingly, the above equations are not tractable when seeking explanatory analytical solutions, but the problem simplifies dramatically for a 1D (Bragg) crystal. Equation (7.12b) therefore simplifies (after some effort) to

$$\frac{1}{\varepsilon_r} \left(-\frac{\partial^2 E_y}{\partial x^2} - \frac{\partial^2 E_y}{\partial z^2} \right) = \left(\frac{\omega}{c}\right)^2 E_y \tag{7.15}$$

where E_y is the electric field on the y-axis of a Cartesian coordinate system [32]. If the wave is a plane wave propagating on the z-axis only then we have

$$\frac{-1}{\varepsilon_r} \frac{d^2 E_y}{dz^2} = \left(\frac{\omega}{c}\right)^2 E_y \tag{7.16}$$

This is a simple second order ordinary differential equation, but unfortunately we must deal with $1/\varepsilon_r$ that is a rectangular periodic function along the z-axis. For simplicity we can represent this as

$$\varepsilon^{-1}(z + n\Lambda) = e^{jnk_z \Lambda} \text{rect}(z) \tag{7.17}$$

which is a square wave.

A full and rigorous analysis of this problem may be found in Joannopoulos et al. [32] but here we will make the following observation; when the propagating wave is well-matched to the crystal structure, the two periodic functions will be closely correlated giving rise to non-propagating standing waves as described above. When the crystal and the wave are less well correlated there will be a propagating wave mode. Using this approach we may, therefore, recast Eq. (7.16) as a straightforward correlation calculation. It is convenient to perform this calculation in k-space.

To do this we must find the k-space equivalents for both the terms on the left-hand side of Eq. (7.16). The rectangular function of Eq. (7.17) Fourier transforms as follows:

$$\varepsilon^{-1} = rect(z) \overset{F}{\to} \text{sinc } k_z \Lambda \tag{7.18}$$

similarly, the second order derivative term Fourier transforms like this:

$$\frac{d^2 E_y}{dz^2} \overset{F}{\to} -k_z^2 E_y \tag{7.19}$$

We can now write down the standard correlation function for Eqs. (7.18) and (7.19), which is

$$\int_{-k/2}^{k/2} (k_z + nK)^2 \sin c(k_z \Lambda) dk_z = \left(\frac{\omega}{c}\right)^2 \tag{7.20}$$

Ordinarily the shift parameter in the quadratic term would only be $(k_z + K)^2$. In this case K is the crystal vector (in 1D), hence by applying Bloch's theorem we may use $(k_z + nK)^2$, so that we find solutions for all the Brillouin zones of the crystal. The full solution for Eq. (7.20) is

$$\frac{\sin k\Lambda - k\Lambda \cos k\Lambda + (2\pi \Lambda n)^2 \left(k\Lambda + \frac{(k\Lambda)^3}{3.3!} + \frac{(k\Lambda)^5}{5.5!} + \cdots\right)}{4\Lambda^3} = \left(\frac{\omega}{c}\right)^2 \tag{7.21}$$

A typical representation of Eq. (7.21) is sketched in Fig. 7.13. The dashed lines show the dispersion relation for freely propagating forward and backward waves. Where solutions (modes) exist they fall inside the free-space dispersion relations. Gaps appear between the crystal modes; these are the band-gaps. Figure 7.13 shows the first 3 positive and negative modes or Brillouin zones. It is widespread practice to redraw, by a process of folding, the zones with $|n| > 1$ inside the first Brillouin zone, thus giving the standard picture of an EBG or EBG dispersion relation. The equivalent such diagram for Fig. 7.13 is shown in Fig. 7.14a. Figure 7.14b shows the same graph when plotted using Eq. (7.21) for the first 5 modes.

Of more relevance to a filter designer is the transmission coefficient of the component as a function of frequency. The first bandgap, as depicted in Fig. 7.14, will in practice lead to a spectrum akin to those shown in Fig. 7.15. Figure 7.15a shows the spectrum for a crystal made of an insulator with different dielectric constant, whereas Fig. 7.15b shows the spectrum for a metal/insulator crystal.

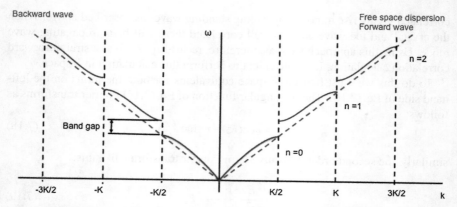

Fig. 7.13 The band diagram for the first three Brillouin zones of a photonic crystal

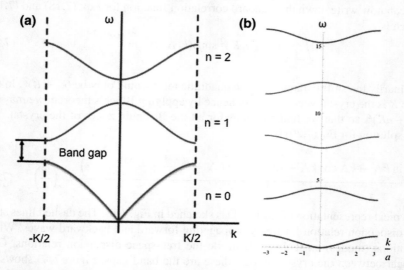

Fig. 7.14 **a** A schematic representation of the first three bands in the Brillouin zone and **b** the results for the first five bands determined using Eq. (7.21)

The basic principle, that of using a synthetic crystal to create an EBG, is the same. However, there is a cut-off at low frequencies due to the finite apertures in the metallic portion of the structure supporting only evanescent modes that are akin to a hollow waveguide in cut-off. This phenomenon gives rise to a narrowband passband resonance that can be immediately exploited, and it has been shown that the position of the resonance may be tuned to realise a tuneable filter.

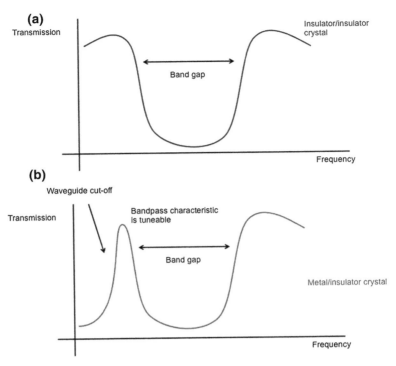

Fig. 7.15 The transmission versus frequency response of **a** and insulator/insulator EBG filter and **b** the response of a metal/insulator filter

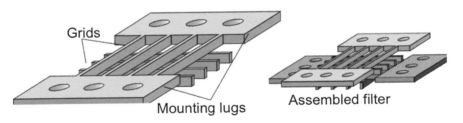

Fig. 7.16 A schematic of a tuneable EBG made using two waffle plates

7.3.2 Realisation of a Tuneable Terahertz Filter

Figure 7.16 shows a schematic of the filter design. It is composed of two metallic "waffle" plates that are milled or etched with gratings on both the front and back faces. The two gratings are at right angles and are etched deeply enough to meet at their deepest point to create an open, but mechanically rigid plate.

Fig. 7.17 Tuning is achieved by a lateral shift of one plate with respect to the other

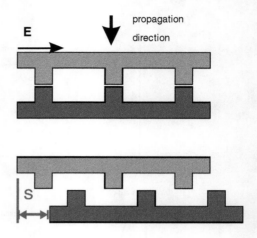

A tuneable EBG filter is made by bringing two plates together (Fig. 7.16). The tuning mechanism is obtained by moving one of the plates as shown in Fig. 7.17. Figure 7.18 shows examples of two different filters. The structure in Fig. 7.18a is made by conventional milling of thin aluminium plates [31]. Mechanical milling in this way is a widely available technique, but is limited in the sizes that can be achieved so the devices work in the millimetre wave regime. In order to make small structures to access higher frequencies it is necessary to use microfabrication techniques [26, 28, 29].

Figure 7.18b shows a structure made using photolithography and deep-reactive ion etching [29]. The device is made from silicon which is amenable to deep etching, and is plated with c. 1 μm of gold so that the whole structure behaves as though it is metallic. Both the plates shown are made by cutting or etching gratings on opposite sides of the sample, but at right angles. As a consequence through plate openings

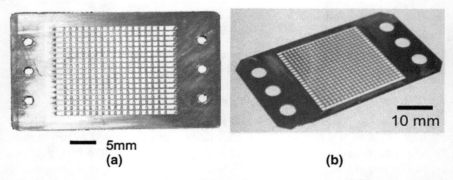

Fig. 7.18 a An aluminium plate for a tunable EBG filter and **b** a gold-plated silicon micromachined plate

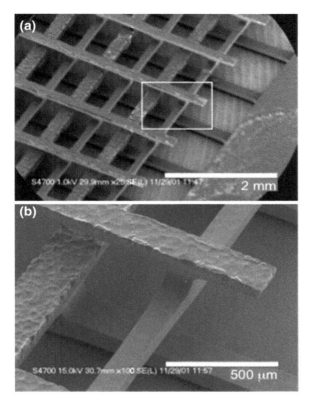

Fig. 7.19 Scanning electron micrographs at two magnifications revealing the detail of an assembled tuneable EBG filter

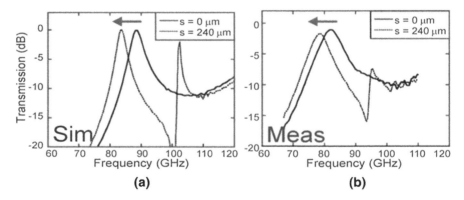

Fig. 7.20 **a** The simulated, and **b** the corresponding experimental response and tuning characteristic, of the aluminium EBG filter

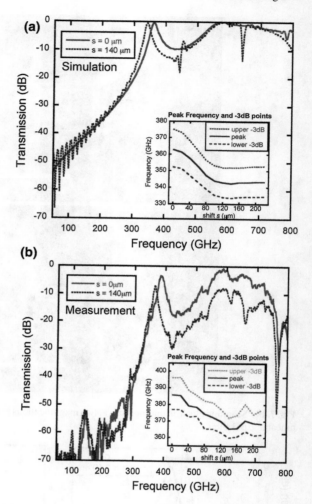

Fig. 7.21 a Simulation of a 350 GHz EBG filter and **b** its experimental characteristics. The inset in **b** shows the passband centre and edge tuning characteristic

are made, but the sample maintains its structure since the gratings are joined where they cross over. Figure 7.19 shows the assembly of two of the silicon plates to form a complete filter. Once assembled the device forms a complex 3D structure that provides the filter function.

Figures 7.20 and 7.21 show the characteristic response and tuning of two filters for W-Band and operation near 350 GHz, respectively. As can be seen, the filters have the expected passband response and exhibit tunability. Detailed investigation of the 350 GHz device demonstrated that a Q of, approximately, 10 is achievable for the passband response, and that the Q is relatively invariant as a function of the centre frequency.

7.4 Metamaterials and Surface Plasmon Resonance

Metamaterials provide a means by which notional synthetic materials can be fabricated so that novel electromagnetic behaviour can be obtained. Metamaterials are composed of insulators and metals. Typically, the excitation frequency will be relatively low such that the metals can be treated, to all intents and purposes, as perfect conductors. Several possibilities for materials have been postulated including negative refractive index that may give rise to phenomena such as invisibility cloaking and the perfect lens [33–38]. More down to earth applications are to use metamaterials to make filters—akin to frequency selective surfaces that are well known—or absorbers that have applications in detector technology [39–44].

A second phenomenon that is well worth considering is that of surface plasmon resonance (SPR). Surface plasmons arise from the rhythmic oscillation of free carriers in a conductor in response to illumination by an electromagnetic field [45, 46]. The relative permittivity is strongly dependent on the frequency of oscillation, and at low frequencies the permittivity of a metal has a large imaginary part that accounts for conductivity hence strong classical absorption. The dielectric function is characterised by a plasma frequency (ω_p), and its behaviour with frequency can be modelled using the Drude relation thus:

$$\varepsilon_1(\omega) = 1 - \frac{\omega_p^2 \tau^2}{1 + \omega^2 \tau^2} \tag{7.22}$$

$$\varepsilon_2(\omega) = \frac{\omega_p^2 \tau}{\omega(1 + \omega^2 \tau^2)} \tag{7.23}$$

where ε_1 and ε_2 are the real and imaginary parts of the dielectric function respectively, and τ is the carrier scattering relaxation time.

The real and imaginary parts of the dielectric function close to resonance in a metal such as gold are shown in Fig. 7.22 [47]. As indicated by the Drude relations, as the excitation frequency of the incident light approaches the plasma resonance, the nature of the material changes dramatically, in the case of a conductor, to become insulating. This phenomenon is widely exploited in the visible spectrum using gold or

Fig. 7.22 The real and imaginary part of the dielectric function for gold near ω_p

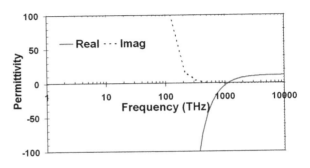

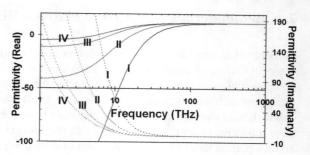

Fig. 7.23 The real and imaginary parts of the dielectric function for silicon near ω_p with various donor concentrations. *I* 5×10^{19} cm^{-3}, *II* 1.6×10^{19} cm^{-3}, *III* 7×10^{18} cm^{-3}, and *IV* 5×10^{18} cm^{-3}. The *solid lines* are for the real part and the *dashed lines* are for the imaginary part [47]

silver as the dielectric, since light passing through a thin film will transfer energy from the magnetic field into its electric field, leading to a strong field enhancement. This enhancement of the electric field may, therefore, be exploited for sensor applications. The reason that metals such as gold and silver show a plasma resonance in the visible (or near visible) spectrum is that they have a very high free-carrier density. The plasma frequency may be calculated using

$$\omega_p^2 = \frac{e^2 N}{\varepsilon_0 m^*} \tag{7.24}$$

where N is the free carrier density, e is the charge on an electron, ε_0 is the dielectric constant of free space and m^* is the effective mass of the carrier. As a consequence SPR can be observed and exploited in the terahertz band by choosing a suitable conductor.

Silicon can be used to make a material exhibiting SPR in the terahertz region, as illustrated in Fig. 7.23 for several different doping concentrations [47, 48]. Using micromachining methods it is, therefore, possible to fabricate free-standing SPR resonators in silicon on a SiO$_2$ membrane. These are shown in Fig. 7.24 for micro-dot resonators, and for micro-ring resonators in Fig. 7.25.

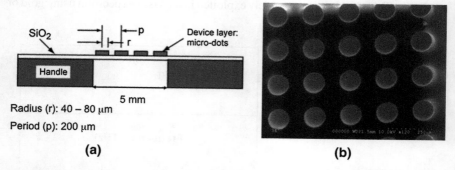

Fig. 7.24 **a** A cross-section through a silicon on insulator wafer with micro-dot resonators on a membrane, and **b** a scanning electron micrograph of the dot array [48]

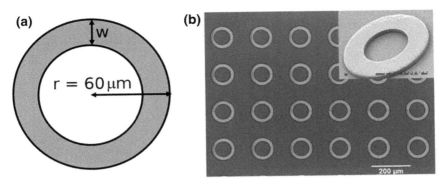

Fig. 7.25 **a** A sketch and **b** optical micrograph of micro-ring SPR terahertz resonators [47]. The inset shows an SEM image of a single micro-ring resonator

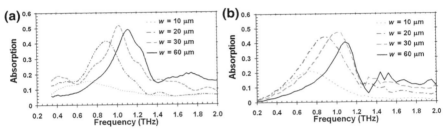

Fig. 7.26 **a** Simulation, and **b** experimental data from the micro-ring resonators of Fig. 7.2 [47]

The critical dimensions for the resonators are not only the resonator radii, but also the grating period, since for the devices to work it is necessary to couple radiation into the resonators. This is most readily achieved using a grating to phase-match the incident beam to the structure, hence exciting the resonant modes. Spectroscopic results, taken using a FTIR, demonstrate the absorption characteristics of silicon micro-ring resonators are shown in Fig. 7.26.

Metamaterials offer a potentially similar route to achieving the same effect, but by exploiting the geometric structure of a metal more fully to attain the resonant characteristic instead of SPR. Figure 7.27 shows a microfabricated absorber structure and the resulting absorption spectra. The structure relies on a continuous metal backplane coated with a thin insulator—in this case spin-cast polyimide. Electromagnetic resonators, in a periodic pattern, are fabricated on top of the polyimide. The structure has a similar characteristic to the SPR device, but is capable of achieving higher absorption; an attribute that may be of use when making sensors.

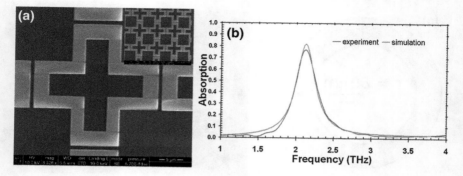

Fig. 7.27 **a** A scanning electron micrograph of a single metamaterial absorber. The inset shows a section of the array. **b** Experimental and simulation data of a THz metamaterial absorber [43]

7.5 Summary

Passive optical components are essential for light control in any optical system. Terahertz technology presents some interesting challenges and opportunities for optical engineering owing to the long wavelength of the radiation. The challenges include the need to access dimensions that are too small for conventional machining in most cases, but too large for most microfabrication technologies. Fortunately, the development of technologies for mainstream micro-electro-mechanical systems (MEMS) such as deep reactive ion etching can be used successfully.

In this chapter we have reviewed a number of optical components that can be micromachined or microfabricated. These include relatively conventional lenses based on Fresnel zone plates. Working at long wavelengths also creates opportunities to investigate novel devices using artificial dielectric methods. In this chapter we have seen how new kinds of variable retarders can be made. The electromagnetic bandgap is a widely known phenomenon and has a practical application in making filters. Using MEMS methods it is possible to make a range of devices, and in this chapter we have reviewed the principle of operation of an EBG filter and the design and implementation of a tuneable device. Finally, we have looked at terahertz absorbing structures. These are particularly applicable to the development of sensors and detectors. Two methods have been presented based on exploitation of surface plasmon resonance and metal structures that borrow from the emerging topic of metamaterials research.

This brief chapter has perhaps only touched upon the surface of a very large topic, and it is to be expected that many new techniques and methods of manufacture will be investigated and implemented in the years ahead.

References

1. E. Hecht, *Optics*, 4th edn. (Addison Wesley, San Francisco, 2002)
2. E.D. Walsby, et al., Fabrication of multilevel silicon diffractive lenses for terahertz frequencies, in Micromachine Technology for Diffractive and Holographic Optics, pp. 79–87 (1999)
3. E.D. Walsby et al., Analysis of silicon terahertz diffractive optics. Curr. Appl. Phys. **4**(2–4), 102–105 (2004)
4. E.D. Walsby et al., Investigation of a THz Fresnel lens. Ultrafast phenomena XIII, pp. 292–294 (2003)
5. E.D. Walsby et al., Multilevel silicon diffractive optics for terahertz waves. J. Vac. Sci. Technol. B **20**(6), 2780–2783 (2002)
6. E.D. Walsby et al., Silicon diffractive optics at THz frequencies, in Diffractive Optics and Micro-Optics, Proceedings Volume, pp. 35–37 (2002)
7. S. Wang et al., Characterization of T-ray binary lenses. Opt. Lett. **27**(13), 1183–1185 (2002)
8. S. Wang, X.C. Zhang, Tomographic imaging with a terahertz binary lens. Appl. Phys. Lett. **82**(12), 1821–1823 (2003)
9. E.D. Walsby et al., Imprinted diffractive optics for terahertz radiation. Opt. Lett. **32**(9), 1141–1143 (2007)
10. E.D. Walsby et al., Fabrication of terahertz holograms. J. Vac. Sci. Technol. B **25**(6), 2329–2332 (2007)
11. M.S. Heimbeck et al., Terahertz digital holography using angular spectrum and dual wavelength reconstruction methods. Opt. Express **19**(10), 9192–9200 (2011)
12. *The Handbook of Optical Constants of Solids*, vol. 3, (Academic Press Limited, London, 1998)
13. E. Wolf, *Principles of optics*. 7 edn. (Cambridge University Press, Cambridge, 1999)
14. P.B. Clapham, M.C. Hutley, Reduction of lens reflection by Moth eye principle. Nature **244**(5414), 281–282 (1973)
15. D.H. Raguin, G.M. Morris, Analysis of antireflection-structured surfaces with continuous one-dimensional surface profiles. Appl. Opt. **32**(14), 2582–2598 (1993)
16. D.H. Raguin, G.M. Morris, Antireflection structured surfaces for the infrared spectral region. App. Opt. **32**(7), 1154–1167 (1993)
17. S.C. Saha et al., Method for vector characterization of polar liquids using frequency-domain spectroscopy. Opt. Lett. **36**(17), 3329–3331 (2011)
18. D.R.S. Cumming, R.J. Blaikie, A variable polarisation compensator using artificial dielectrics. Opt. Commun. **163**(4–6), 164–168 (1999)
19. T.D. Drysdale et al., Variable polarisation compensator using artificial dielectrics for millimetre and submillimetre waves. Electron. Lett. **37**(3), 149–150 (2001)
20. T.D. Drysdale et al., Artificial dielectric devices for variable polarization compensation at millimeter and submillimeter wavelengths. IEEE Trans. Antennas Propag. **51**(11), 3072–3079 (2003)
21. S.A. Jewell et al., Tuneable fabry-perot etalon for terahertz radiation. New J. Phys. **10**, 033012 (2008)
22. C.J.E. Straatsma, A.Y. Elezzabi, A dual-mode terahertz filter based on a metallic resonator design. J. Infrared Millimeter Terahertz Waves **32**(11), 1299–1306 (2011)
23. E.S. Lee et al., Terahertz notch and low-pass filters based on band gaps properties by using metal slits in tapered parallel-plate waveguides. Opt. Express **19**(16), 14852–14859 (2011)
24. N. Jin, J.-S. Li, Terahertz wave bandpass filter based on metamaterials. Microw. Opt. Technol. Lett. **53**(8), 1858–1860 (2011)
25. I.S. Gregory et al., Multi-channel homodyne detection of continuous-wave terahertz radiation. Appl. Phys. Lett. **87**(3), 034106 (2005)
26. T.D. Drysdale et al., Terahertz tuneable filters made by self-releasing deep dry etch process. Microelectron. Eng. **73–4**, 441–446 (2004)
27. T.D. Drysdale et al., Transmittance of a tunable filter at terahertz frequencies. Appl. Phys. Lett. **85**(22), 5173–5175 (2004)

28. R.J. Blaikie et al., Wide-field-of-view photonic bandgap filters micromachined from silicon. Microelectron. Eng. **73–4**, 357–361 (2004)
29. T.D. Drysdale et al., Metallic tunable photonic crystal filter for terahertz frequencies. J. Vac. Sci. Technol. B **21**(6), 2878–2882 (2003)
30. T.D. Drysdale, R.J. Blaikie, D.R.S. Cumming, A tunable photonic crystal filter for terahertz frequency applications, in *Terahertz for Military and Security Applications*, pp. 89–97 (2003)
31. J.D. Joannopoulos, S.G. Johnson, J.N. Winn, R.D. Meade, *Photonic Crystals: Molding the Flow of Light*, 2nd edn. (Princeton University Press, Princeton, 2008)
32. T.D. Drysdale, R.J. Blaikie, D.R.S. Cumming, Calculated and measured transmittance of a tunable metallic photonic crystal filter for terahertz frequencies. Appl. Phys. Lett. **83**(26), 5362–5364 (2003)
33. N. Fang et al., Sub-diffraction-limited optical imaging with a silver superlens. Science **308**(5721), 534–537 (2005)
34. J.B. Pendry, Negative refraction makes a perfect lens. Phys. Rev. Lett. **85**(18), 3966–3969 (2000)
35. X. Zhang, Z.W. Liu, Superlenses to overcome the diffraction limit. Nat. Mater. **7**(6), 435–441 (2008)
36. W.S. Cai et al., Optical cloaking with metamaterials. Nat. Photonics **1**(4), 224–227 (2007)
37. D. Schurig et al., Metamaterial electromagnetic cloak at microwave frequencies. Science **314**(5801), 977–980 (2006)
38. J. Valentine et al., An optical cloak made of dielectrics. Nat. Mater. **8**(7), 568–571 (2009)
39. M.J. Dicken et al., Frequency tunable near-infrared metamaterials based on VO_2 phase transition. Opt. Express **17**(20), 18330–18339 (2009)
40. N.I. Landy et al., Perfect metamaterial absorber. Phys. Rev. Lett. **100**(20), 207402 (2008)
41. H. Tao et al., A metamaterial absorber for the terahertz regime: design, fabrication and characterization. Opt. Express **16**(10), 7181–7188 (2008)
42. J. Grant et al., Polarization insensitive, broadband terahertz metamaterial absorber. Opt. Lett. **36**(17), 3476–3478 (2011)
43. J. Grant et al., Polarization insensitive terahertz metamaterial absorber. Opt. Lett. **36**(8), 1524–1526 (2011)
44. D.M. Wu et al., Terahertz plasmonic high pass filter. Appl. Phys. Lett. **83**(1), 201–203 (2003)
45. K.L. Kelly et al., The optical properties of metal nanoparticles: the influence of size, shape, and dielectric environment. J. Phys. Chem. B **107**(3), 668–677 (2003)
46. S. Maier, Plasmonics: fundamentals and applications (Springer, Berlin, 2007)
47. J. Grant et al., Terahertz surface plasmon resonance of periodic silicon micro-dot arrays, in *2010 IEEE Photonics Society Winter Topicals Meeting Series*, 2010. pp. 34–35
48. J. Grant et al., Terahertz localized surface plasmon resonance of periodic silicon microring arrays. J. Appl. Phys. **109**(5), 054903 (2011)

Chapter 8
Principles and Applications of THz Time Domain Spectroscopy

J.-F. Roux, F. Garet and J.-L. Coutaz

Abstract Terahertz Time-Domain Spectroscopy (THz-TDS) is one of the most powerful techniques to characterise the far infrared response of materials and devices. Based on the use of short electromagnetic waveforms generated by rectifying femtosecond optical pulses delivered by a mode-locked laser, it allows, within one single experiment, the determination of the optical constants of materials over a wide frequency spectrum ranging from 100 GHz up to several THz. Thanks to its large dynamics and its ability to give both amplitude and phase of the transmitted, reflected or scattered THz fields, this technique addresses a large panel of scientific studies. In this chapter, one introduces the principles of THz-TDS. First, the characteristics and performances of a classical set-up are described, then the extraction of the refractive index and absorption coefficient of the studied device from its complex experimental transmission or reflection coefficients are explained. As a demonstration of the power of this technique, one presents some results of thin film spectroscopy and recent studies of left-handed metamaterials. Finally, a detailed analysis of the precision of determined optical constants is given. One will treat not only the classical transmission scheme, but also the reflection regime that is mandatory in the study of strongly absorbing media.

Keywords Terahertz time-domain spectroscopy · THz-TDS · Refractive index · Absorption coefficient · Left-handed metamaterials · Photoconductive antenna

8.1 Introduction

Far infrared properties of materials have been studied for more than one century, since the pioneering works of J. Bose, H. Rubens and E. F. Nichols [1, 2]. Until the 1980s, the most powerful tool was the Fourier Transform Infrared (FTIR) spectrometer based

J.-F. Roux (✉) · F. Garet · J.-L. Coutaz
Laboratoire IMEP-LAHC–UMR CNRS 5130, Université de Savoie, rue du Lac de la Thuile, 73376 Le Bourget du Lac, France
e-mail: Jean-Francois.Roux@univ-savoie.fr

M. Perenzoni and D. J. Paul (eds.), *Physics and Applications of Terahertz Radiation*, Springer Series in Optical Sciences 173, DOI: 10.1007/978-94-007-3837-9_8, © Springer Science+Business Media Dordrecht 2013

on a black body source, a Michelson interferometer and a bolometer [3]. During the 1970s, the emergence of mode-locked lasers, delivering picosecond optical pulses, has allowed researchers to produce short electrical pulses by rectifying the optical pulse envelop by means of non-linear effects in crystals [4] or in the fast response of semiconductor photoswitches (PSW) [5]. The initial research was focused on the optoelectronic generation of stable and short electrical pulses propagating along transmission lines and a lot of work was initialised in the field of fast optoelectronics thanks to the ability of accurately measuring such pulses by electro-optic sampling [6]. At the same time, it was demonstrated that the generation of short electrical pulses in PSWs also led to the radiation of short electromagnetic pulses propagating in free space: the so-called "Auston switch" acting in fact as a Hertzian dipole [7]. Thanks to the use of femtosecond mode-locked lasers, the radiated spectrum spans from 0.1 to several THz. The first studies of materials or gases using such pulses were published in the late 1980s, while the experimental technique started to be described as "Terahertz Time-Domain Spectroscopy" (THz-TDS) [8, 9].

Since these pioneering works, this technique has been widely used to study the THz response of materials and devices with applications in the fields of physics, chemistry, biology, medicine and technology [10]. Now, several TDS systems are commercially available, some of them being directly derived from laboratory set-ups while others are compact and integrated with more sophisticated human machine interface. The main advantage of TDS relies on the one-measurement determination of the broadband response (up to few THz) of a device (sample) as well as the high stability of the mode-locked lasers, which lead to very good signal-to-noise ratio (SNR). Moreover, the SNR is reinforced by the time-windowing approach that eliminates a large part of the noise. The versatility of the THz-TDS set-ups based on quasi-optical elements offers a large variety of configurations and thus of measurements. Actually, THz-TDS allows one to measure simultaneously the amplitude and phase of the THz signals that are transmitted, reflected, deflected or scattered by the sample, without or with loss of coherence. The sample can be a simple slab of matter or a more complex device, such as THz waveguides, antennae, filters... Usually, THz-TDS measurements result in determining the spectra of the refractive index n and of the absorption coefficient α of samples, but the sample thickness as well as scattering or diffraction parameters can also be extracted. Using the same experimental set-up, the study can be totally performed in the time domain: thanks to the very short THz pulse duration, high resolution tomography can be realised [11].

In this chapter, one will present the principles of THz-TDS. Then one will describe THz-TDS set-ups with emphasis on their performances in terms of bandwidth, SNR and dynamics. One will then explain in detail how to extract the refractive index and coefficient of absorption of materials from THz-TDS data. As an example of the variety of allowed measurements, one will present some results on thin films characterisation and studies of metamaterials showing left-handed behaviour in the THz frequency range. Finally, one will conclude with some developments about uncertainties and accuracy of spectroscopy measurements done either in transmission or reflection.

Fig. 8.1 The principle of
the differential time-domain
measurement

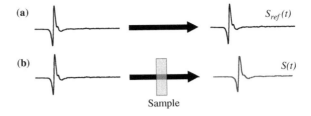

8.2 Principles of THz-TDS

Basically, THz-TDS is a technique based on the measurement of a short electromagnetic pulse that is transmitted, reflected or scattered by a sample. Here, for sake of simplicity, one will describe the experiment performed in a transmission scheme. One assumes that such a short pulse has been generated and can be detected in a synchronous manner. In a first step, the reference waveform $S_{ref}(t)$ that has been transmitted in air is recorded versus the time delay t between emission and reception (Fig. 8.1a). Then, keeping the same time origin, the signal $S(t)$ transmitted by the sample under test is recorded (Fig. 8.1b). This signal is fundamentally affected by the sample: its amplitude is reduced because of Fresnel reflection at each interface and due to possible absorption or scattering in the material. Moreover, the signal is delayed because of the time-of-flight across the sample. As it is actually difficult to model the material response in the time domain, one prefers to calculate the complex transfer function $T(\omega)$ of the sample in the frequency domain.

To get $T(\omega)$, one calculates the Fast Fourier Transform (FFT) of both $S(t)$ and $S_{ref}(t)$, respectively $S(\omega)$ and $S_{ref}(\omega)$ (Fig. 8.2). Assuming the shape and amplitude of the input pulse remain constant during the two-measurements process, the complex transfer function $T(\omega)$ of the sample is simply given by:

$$T(\omega) = t(\omega) e^{j\varphi_t(\omega)} = \frac{S(\omega)}{S_{ref}(\omega)}. \tag{8.1}$$

As $T(\omega)$ is a function of the dielectric response of the sample, one can determine the complex refractive index $\tilde{n} = n - j\kappa$ of the sample by solving the inverse problem, i.e. by finding the value of $\tilde{n}$ for which experimental and calculated $T(\omega)$ equalise. The procedure for the most common case, i.e. when the sample is a homogeneous slab with flat and parallel faces will be explained precisely later. In any case, the expression of the complex transmission coefficient must be known to perform the extraction procedure once this coefficient has been measured by THz-TDS.

So the TDS experiment is based on the ability of producing and detecting coherently short electromagnetic pulses that interact with the sample and on the possibility of knowing analytically or by simulation the sample response.

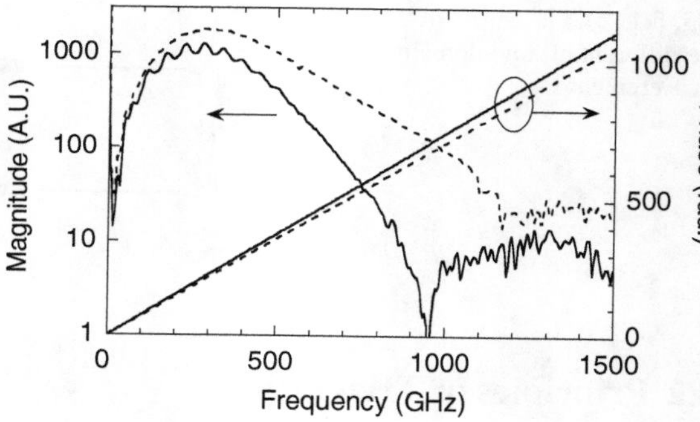

Fig. 8.2 The frequency spectra $S_{ref}(\omega)$ (*dashed line*) and $S(\omega)$ (*continuous line*) of the reference and the signal transmitted by a 1.065-mm thick slab of BK7 glass

Fig. 8.3 Basic scheme of a pulsed laser-assisted generation and detection of THz signals in PSWs. Here, the THz beam is made almost parallel by means of two hyper-hemispherical lenses

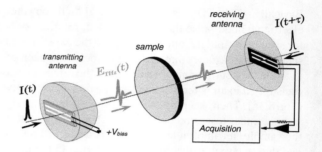

8.3 Generation and Detection of Short Electromagnetic Pulses

As already mentioned in the introduction, the generation of electromagnetic pulses by rectification of sub-picosecond optical pulses has been first demonstrated in non-linear crystals. More rigorously, this process must be described as the frequency difference of the spectral components of the broadband optical pulse. It has a weak efficiency but allows one to reach very high THz frequencies and very strong THz electrical fields [12]. Similarly, the electro-optic Pockels effect can be used to detect fast THz transients in a thin non-linear crystal such as ZnTe, which minimises group velocity dispersion effects between the THz pulse and the optical probe [13]. Here, one will focus our attention on the generation and detection of the electrical transients in PSW made from ultrafast semiconductor materials such as LTG-GaAs [14]. Actually, these devices only require a few mW of laser power to generate broadband THz pulses with an average power of a few tens of nW and a dynamic range (DR) better than 60 dB. The generation of THz pulses by photo-switching in semiconducting antennae can be described almost analytically within some basic assumptions, and thus these THz antennae can be optimised with regard to the parameters of the semiconductor material [15].

As shown in Fig. 8.3, the transmitting antenna consists usually in two parallel metallic strips biased by a constant voltage V_{bias} and deposited onto a semiconductor layer of high resistivity, exhibiting good carrier mobility and a carrier lifetime in the ps or sub-ps range. The optical beam is focused down to a small spot illuminating the absorbing semiconductor layer near the anode location or at a place (called "the gap") where the distance in between the metallic lines is reduced down to a few μm. When absorbed by the semiconductor, the ultra-short optical pulse $I(t)$ generates photocarriers almost instantaneously, giving rise to a burst of current $i(t)$ whose temporal shape depends mostly on the laser pulse duration and the free electron lifetime τ_e (because of their reduced mobility, holes contribution can be neglected in a first-order analysis). In the case of long carrier lifetime, i.e. larger than some ps, the temporal shape of the current burst depends also on the carrier transit time in between the electrodes and on the RC constant time of the antenna circuit. The photocarrier density in the PSW at time t is the sum of all the photocarriers generated by the laser pulse $I(t - t')$ at the preceding times t', so assuming that the carriers relax exponentially with a decay time τ_e, the generated current for a Gaussian pumping optical pulse writes as:

$$i(t) \propto \int_{-\infty}^{t} P_o \exp\left[-\left(\frac{t - t'}{\Delta t}\right)^2\right] \exp\left(-\frac{t'}{\tau_e}\right) \mu E_{\text{Bias}} dt' \qquad (8.2)$$

where P_o is the incident optical power, Δt is the duration of the optical pulse, μ is the electron mobility (here considered as constant) and E_{bias} is the biasing DC field. Let us notice that Eq. (8.2) is the convolution product of the laser pulse and the photocarrier density. The PSW, whose geometrical parameters are smaller than the involved THz wavelengths, may be considered as an electrical dipole, whose radiated far field E_{THz} is proportional to the time derivative of the current burst $i(t)$:

$$E_{\text{THz}}(t) \propto \frac{\partial i(t)}{\partial t}. \qquad (8.3)$$

So in the linear emitting regime of the PSW, the amplitude of the THz field is a linear function of both the optical power and the biasing electrical field. For large values of these excitation parameters, different physical phenomena like saturation of the carrier drift velocities or screening of the biasing field limit the THz amplitude [16]. The calculation of the emitted field amplitude should take into account the impedance of the antenna electrical circuit acting together with the equivalent impedance of the focusing hemispherical lens (see Fig. 8.3). This latter, made from high-resistivity silicon, is optimising the impedance matching in between the antenna substrate ($\varepsilon \approx 11.4$) and air ($\varepsilon = 1$).

At the receiver, a time-delayed replica of the laser pulse $I(t + \tau)$ triggers a similar ultrafast PSW. This time delay τ is adjusted thanks to an optical delay line (not shown in Fig. 8.3). In the receiver, the impinging THz field collected by the metallic antenna accelerates the carriers photo-generated by the delayed optical pulse. Assuming, as one did for the emitter, an exponential decay for the photocarriers population, the current delivered by the receiver and integrated by a slow electronics apparatus is given by:

$$i_{\text{det}}(\tau) \propto \int_{-\infty}^{+\infty} E_{\text{THz}}(t - \tau) \left[\int_{-\infty}^{t} P_o \exp\left[-\left(\frac{t - t'}{\Delta t}\right)^2 \right] \exp\left(\frac{-t'}{\tau_e}\right) dt' \right] dt.$$

$$(8.4)$$

If the optical pulse duration is very short (much shorter than the THz pulse and the carrier lifetime), two types of detectors may be identified, depending on the value of τ_e [17]. For carrier lifetime much shorter than the THz pulse duration, the PSW response to the THz field is almost instantaneous and the PSW acts as a "direct sampling detector", with $i_{\text{det}}(\tau) \propto E_{\text{THz}}(\tau)$. On the other hand, for PSW with very long carrier lifetime, the THz field is fully time integrated by the long living photocarriers and then the PSW acts as an "integrating detector". It has to be pointed out that the performances of both kinds of receivers in terms of bandwidth are roughly similar because, as for the emitter, the THz bandwidth is mainly determined by the rate at which the carriers are photogenerated in the detector, i.e. by the laser pulse duration. But, as explained in a forthcoming section devoted to noise, PSWs with long carrier lifetime present a greater sensitivity to sub-THz frequencies, but exhibit a poorer SNR for the highest THz frequencies, as they also integrate a large amount of white noise.

Usually, the THz-induced current in the receiver returns to zero before a second optical pulse of the laser comb impinges the system. Therefore, the detected signal may be integrated over a large number of pulses with, for example, a lock-in amplifier. As already mentioned, the optical delay line enables the ability to scan the time delay τ and thus to record the signal over a time window T_{window}. The choice of T_{window} is of particular importance because it has to be larger than the THz pulse at the receiver, it defines the spectral resolution of the system, and it influences the SNR of the measurement.

8.4 Characteristics of a TDS Set-up

A common TDS system consists of a mode-locked fs laser, two PSWs working respectively as THz transmitter and receiver, a quasi-optical system that shapes the THz beam and directs it from the emitter onto the receiver through the sample, an optical delay line allowing sampling of the signal and a detection electronics system. In this section, one will investigate how the characteristics of these components affect the overall performances of the complete TDS system.

8.4.1 Bandwidth

Many factors impact the overall bandwidth of the system. Generally, commercial systems deliver typically signals from 0.1 up to 3–6 THz. Ultra broadband systems, especially built in research laboratories with PSWs, electro-optic antennae or even optical breaking in air excited by laser pulses shorter than 10 fs, reach several tens

Fig. 8.4 The generated THz spectra calculated from (2) with $\tau_e = 500$ fs for different laser pulse duration Δt. The inset shows the spectra obtained for $\Delta t = 100$ fs and $\tau_e = 500$ fs (*solid line*) or $\tau_e = 2$ ps (*dotted line*)

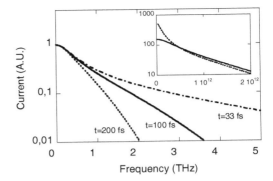

of THz up to more than 200 THz [18]. In such ultra broadband systems, extreme care should be taken concerning dispersion and focusing of the highest frequencies onto the detector within a spatial precision of a few μm. In addition, most of the semiconductors or the non-linear crystals employed in such systems show strong optical phonon absorption bands, in which the THz signal vanishes.

In the case of PSW antennae, the absolute maximum generated bandwidth is defined through Eq. (8.2). Figure 8.4 shows that in this equation, the shortness of the laser pulse width is of major importance to produce a broadband spectrum. On the contrary, the carrier lifetime in the detector and transmitter only slightly affects the power spectral density (PSD) as shown in the inset.

8.4.2 Frequency Resolution

The temporal recorded THz waveform is given by relation (2). This waveform is then Fourier transformed as follows:

$$S_{\text{meas}}(\omega) = \frac{1}{\sqrt{2\pi}} \int_0^{T_{\text{window}}} i_{\text{det}}(\tau) e^{-j\omega\tau} d\tau = \frac{1}{\sqrt{2\pi}} \int_{-\infty}^{+\infty} G(\tau, T_{\text{window}}) i_{\text{det}}(\tau) e^{-j\omega\tau} d\tau,$$

(8.5)

where T_{window} is the scanning window in which the signal is recorded, and $G(\tau, T_{\text{window}})$ is a rectangle function that is equal to 1 for $0 < \tau < T_{\text{window}}$ and 0 elsewhere. The theory of the Fourier transform leads to:

$$S_{\text{meas}}(\omega) = \sqrt{2\pi} S(\omega) \otimes G(\omega, T_{\text{window}}) \propto T_{\text{window}} \ \ S(\omega) \otimes \text{sinc}\left(\frac{T_{\text{window}}\omega}{2}\right),$$

(8.6)

where $\otimes$ is for the convolution product, $G(\omega, T_{\text{window}}) \propto \text{sinc}(T_{\text{window}}\omega/2)$ is the Fourier transform of $G(\tau, T_{\text{window}})$ and $S(\omega)$ is the Fourier transform of the actual signal. Thus, the measured spectrum is equal to the actual one convoluted by a sinc function. The experimental frequency resolution of the system is equal to the spectral width of the sinc function, i.e. $\Delta f = 1/T_{window}$. As an example, typically, THz-TDS systems employ a 5-cm long delay line, leading to $T_{window} = 300$ ps and $\Delta f = 3$ GHz. Longer delay lines require a perfect alignment of the laser beams that should strike the PSWs within a 1-μm precision, as well as an accurate control of the laser power stability. Better resolution is possible by using the mixing of two mode-locked lasers whose repetition rate is slightly different. With such a TDS set-up based on asynchronous optical sampling (ASOPS), resolution as low as 1 GHz can be reached together with the advantage of kHz scanning rates [19].

8.4.3 Spatial Resolution

Spatial resolution in regular TDS systems is ruled by the diffraction limit of the focused THz beam. A 0.1–6 THz broadband spectrum covers the wavelength range of 0.05 to 3 mm. Assuming that the THz beam is focused by a lens with a numerical aperture $D_{diam}/f = 1$, where D_{diam} and f are respectively the diameter and the focal length of the lens, the diameter of the limiting Airy spot, namely $\phi_{airy} = 2.44 \, \lambda f/D_{diam}$, is frequency dependent and varies approximately from 0.12 to a 7.5 mm. Numerous researches have been performed in near-field THz microscopy to achieve a better resolution. For instance, real-time recording of mm^2-images with tens of μm resolution has been demonstrated recently [20].

8.4.4 Signal-to-Noise Ratio of a TDS Set-up

The absolute limits of a TDS set-up are actually limited by its performances in terms of dynamics of the measurement and SNR. Let us recall that the SNR and the DR are defined by:

$$\text{SNR}\,(\omega) = \frac{S\,(\omega)}{\sigma\,(\omega)} \quad \text{and} \quad DR = \frac{S\,(\omega)}{S_{min}\,(\omega)}, \tag{8.7}$$

where $S(\omega)$ is the actual THz amplitude at angular frequency ω, $\sigma(\omega)$ is its standard deviation and $S_{min}(\omega)$ is the minimum measurable signal at ω. If $\sigma(\omega)$ is signal independent, then SNR and DR are equal.

The standard deviation $\sigma(\omega)$ originates from noise. Let us study the sources of noise in a TDS set-up. In a general way, one can identify three classes of noise in such an experiment gathering a transmitter, a receiver (both being activated by a laser source) and a readout electronic apparatus, namely:

– Relative Intensity Noise (RIN) of the THz source,
– Shot noise in the detector,
– Johnson noise in detection and thermal background noise [21].

The amplitude of the THz field varies linearly with the optical power P_o [Eqs. (8.2) and (8.3)], hence the RIN of the THz emitter depends mostly on the RIN of the laser (here the shot noise contribution is negligible). As a consequence, the SNR of the

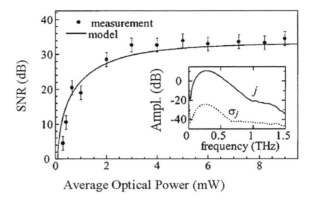

Fig. 8.5 The experimental influence of the RIN noise in a TDS experiment (points) and a comparison (*continuous line*) with Eq. (8.7). The chosen frequency—300 GHz—corresponds to the maximum of the signal spectral density (*j*) as shown in the inset. One can see that noise PSD σ_j is also maximal at this frequency

experiment is depending on the optical power. This is experimentally confirmed in Fig. 8.5 [15] where the SNR at 300 GHz is plotted versus the average optical power P_o focused onto the PSW emitter. The data points are fitted with a simple model neglecting the shot noise in the detector:

$$\text{SNR} = \frac{P_o}{\sqrt{P_o^2 \times \text{RIN} + C}}. \tag{8.8}$$

At small excitation levels, the white noise term C is dominant and the SNR increases linearly with the optical power. This dependency saturates around 4 mW and the RIN of the THz amplitude, directly linked to the RIN of the laser, becomes the main contribution of noise over 300 GHz. Consequently, as shown in the inset of Fig. 8.5, the noise PSD is proportional to the signal up to 600 GHz. For higher frequencies, the THz power decreases due to a weaker generation efficiency and the noise become independent of the THz signal level. Thus, above 600 GHz, the SNR and the DR are equal, whereas the DR is larger than the SNR at smaller frequencies.

The design of the detector has also a strong impact on the DR of the TDS set-up. The important parameters are the spectral gain of the metallic antenna collecting the THz field, the electrical properties of the semiconductor (dark resistivity, carrier mobility and lifetime) and the excitation conditions (photo-carriers density). As an example, one plots in Fig. 8.6 the DR of a conventional set-up (with PSWs used as emitter and receiver) for various carrier lifetimes in the receiver, following the analytical model developed in [15]. Basically, the results show that the larger DR is obtained for rather long ps carrier lifetimes at the cost of a rapid decay above 500 GHz, whereas the shortest lifetimes give more homogeneous, but lower, DR. This is of major importance for broadband TDS set-ups, especially over a couple of THz where the thermal noise dominates because of the low THz signal level. This may result from the noise integration due to the long carrier lifetime, as reported experimentally in [17, 22], and in such a case, one has to select PSWs with the faster response to increase the SNR and the DR of the set-up.

Fig. 8.6 Influence of the carrier lifetime (ranging from 0.125 to 2 ps) in the transmitter and receiver PSWs on the relative dynamics (or DR) of the TDS set-up

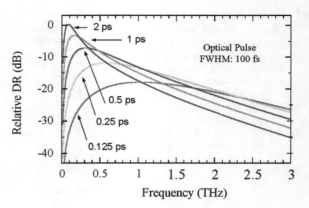

A deeper insight in the physical origin of the noise allows us to derive the dependence of the noise on the signal strength at the detector location. Let us call σ_E the noise of the emitter that has been described above. The THz signal, as well as this noise, is transmitted (reflected) towards the receiver. At the receiver, the THz photons give rise to a shot noise which is signal dependant ($= 2qs\Delta f$, where s is the detected THz signal magnitude, q the charge of the carriers and Δf the detection bandwidth). The detection electronics show a signal-independent noise σ_D (Johnson noise, amplification noise...). Thus the total noise exhibits a quadratic dependence with the transmission (reflection) coefficient $\tau(\rho)$:

$$\sigma_S^2 = x^2\sigma_E^2 + 2qs\Delta f + \sigma_D^2 \ (x = \tau \text{ or } \rho) \tag{8.9}$$

This expression that takes into account the presence of the sample in the analysis of the SNR will be useful while considering the accuracy of the spectroscopy results as it will be developed in the last section of this chapter.

8.4.5 Typical Signals and Set-ups

Here one gives an example of typical signals detected in a THz-TDS experimental set-up based on a mode-locked Ti:Sa laser oscillator delivering 60-fs FWHM duration pulses at a repetition rate of 75 MHz. Both emitter and receiver are LTG-GaAs PSWs (500 fs electron trapping time) with a Grischkowsky-type switch made of evaporated gold strips. The width of the switch gap is about 5-μm and the excitation average power is ranging from 5 to 10 mW. The bias voltage of the emitter is 20 V. The coupling of the PSW to free space is optimised by means of high-resistivity silicon hyper-hemispherical lenses attached onto the switches substrate (Fig. 8.7). For transmission measurements, four gold-plated parabolic mirrors shape the THz beam. For reflection TDS at normal incidence, the THz beam is divided into two

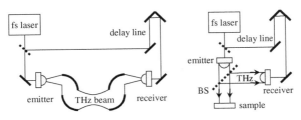

Fig. 8.7 Transmission (*left*) and reflection (*right*) THz-TDS set-ups. Here, reflection occurs under normal incidence. BS is for beam splitter

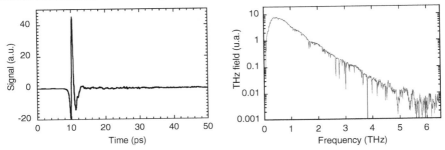

Fig. 8.8 The recorded THz waveform (*left*) and its spectrum (amplitude) calculated by FFT (*right*)

parts by a THz beam splitter. In both cases, the time equivalent sampling detection of the THz pulse is performed thanks to a mechanical delay line whose displacement length allows for a maximum frequency resolution of 3 GHz.

As observed in Fig. 8.8, the duration of the THz pulse detected in transmission is shorter than 1 ps and the FFT spectrum bandwidth is extended to over 5 THz where the signal becomes weaker than the noise. The maximal DR is about 63 dB at 500 GHz. The spectrum shows some dark lines due to residual water vapour absorption in the experimental chamber. This demonstrates the extreme sensitivity of THz waves to water vapour absorption, as air in this chamber has been dried down to 10 % of water vapour. Measurements can also be performed in nitrogen atmosphere in a purged system.

8.5 Principles for Extraction of Optical Parameters from THz-TDS Measurements

THz-TDS set-ups are mainly dedicated to spectroscopy measurements because of the possibility of determining the optical parameters (refractive index n and absorption α) of a material on a very broad frequency range without using Kramers-Kronig analysis. Indeed, both the amplitude and phase of the transmission (reflection) coefficient of a sample are measured. This complex value permits one to obtain the optical parameters of the sample material by solving the inverse electromagnetic problem.

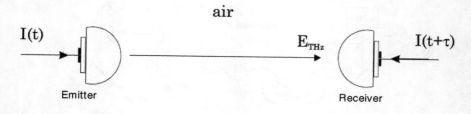

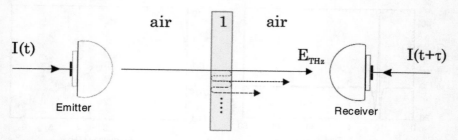

Fig. 8.9 The principle of a THz-TDS measurement in transmission. A first reference signal $S_{ref}(t)$ is measured by direct transmission in air (*upper*). The signal transmitted by the sample $S(t)$ is measured together with (or without) all its temporal replica due to internal reflection (*bottom*)

In the case of transmission, the principle and the different stages of the method are the following (Fig. 8.9):

– Measurement of the THz signals $S(t)$ and $S_{ref}(t)$ respectively obtained with and without the sample in between the emitter and the receiver.
– Calculation of the complex transmission of the sample by calculating the ratio of the complex FFT spectra of the temporal waveforms $S(t)$ and $S_{ref}(t)$.
– Solving of the inverse electromagnetic problem to get n and α.

For opaque or highly absorbing samples, it is preferable to employ a reflection THz-TDS set-up. Let us note that in this case, the reference signal is obtained by putting a perfect mirror at the location of the sample. As the phase of the signal has to be known for the extraction procedure, positioning the mirror with a great accuracy is compulsory. Poor alignment leads to strong errors on n and α. Also, for achieving normal incidence, a THz beam splitter is required. It could be a thin pellicle beam splitter or a thick transparent slab like quartz or high-resistivity silicon wafers, but in any case, it divides the signal reaching the receiver by a factor of at least four.

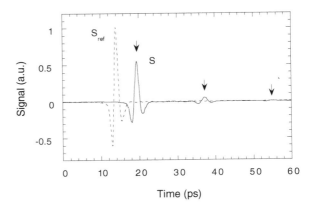

Fig. 8.10 Experimental time-domain signals: reference (*dashed line*) and signal transmitted through a 1.065-mm thick BK7 sample

8.5.1 Temporal Measurements

Two temporal measurements of the THz signal are needed: one $S(t)$ with the sample and a reference one $S_{ref}(t)$ without it. In both cases of transmission and reflection measurements, the reference signal presents a single pulse, whereas the reflected or transmitted signal shows multiple pulses corresponding to echoes in the sample (Fig. 8.9). In the case of optically thick sample (Fig. 8.10), the temporal echoes are sufficiently separated to permit time windowing whereas, for thinner samples, the different echoes are superimposed making inapplicable the time-windowing procedure. Therefore, the complete signal must be recorded in view of the spectroscopy purpose.

The experimental complex transmission coefficient $T_{meas}(\omega)$ of the sample is equal to the ratio of the spectra $S_{ref}(\omega)$ and $S(\omega)$ of the two temporal measurements $S_{ref}(t)$ and $S(t)$. As the phases of the spectra are determined through the time delay of the temporal measurements, the two measurements (reference and signal) must absolutely be recorded with the same time origin. Here the phase and the amplitude of the signals are easily obtained because, as shown in Eq. (8.4), the receiver is sensitive to the field of the THz beam and not to its energy (or intensity), as it is for most of the detectors.

8.5.2 Spectroscopy from Transmission Measurements

Principle

Transmission spectroscopy is mainly used to determine the optical parameters of a material because of its reliability, precision and ease. Using the method described before, the complex transmission coefficient of a sample can be measured. To get the optical parameters, one needs to know the electromagnetic response of the sample in order to solve this inverse problem. The most common case that we develop here consists of a parallel slab of homogeneous material. One is assuming that the sample is isotropic, free of surface charges and currents, and that the THz amplitude is small, so that the electromagnetic response is linear. For a parallel THz beam, the complex transmission coefficient of such a sample depends on both its thickness d and its complex refractive index of the material $\tilde{n} = n - j\kappa$ and can be written under normal incidence as:

$$T_{\text{theo}}(\omega) = te^{j\varphi_t} = \frac{4\tilde{n}}{(\tilde{n}+1)^2} \exp\left(-j(\tilde{n}-1)\frac{\omega}{c}d\right) FP(\omega), \qquad (8.10)$$

where c is the light velocity in vacuum, $\omega = 2\pi f$ is the angular frequency and:

$$FP(\omega) = \frac{1}{1 - \left(\frac{\tilde{n}-1}{\tilde{n}+1}\right)^2 \exp\left(-2j\tilde{n}\frac{\omega}{c}d\right)} \qquad (8.11)$$

$FP(\omega)$ corresponds to the contribution of all the multiple echoes in the sample, in other words it depicts the Fabry-Pérot resonances. The absorption α of the material is proportional to κ, the imaginary part of $\tilde{n}$:

$$\alpha\left(m^{-1}\right) = \frac{4\pi f \kappa}{c}. \qquad (8.12)$$

The complex refractive index $\tilde{n}$ of the material is obtained solving the inverse electromagnetic problem $T_{\text{meas}}(\omega) - T_{\text{theo}}(\omega) = 0$ where $T_{\text{meas}}(\omega)$ and $T_{\text{theo}}(\omega)$ are complex. Thus, one has to find the zeros of this equation in the complex plane. It is not an obvious problem because of the oscillatory behaviour of $T_{\text{theo}}(\omega)$ through the complex exponential in Eq. (8.10). Nevertheless, this equation can be numerically solved thanks to an error function that allows us to get rid of the oscillations, as one has proposed in [23]. Later on, this method has been numerically improved as for instance in [24]. Figure 8.11 shows the determined refractive index and the absorption of a LaAlO$_3$ sample from 0.1 to 2.3 THz. Generally, the inverse problem is solved with the slab thickness d as a known parameter. However, let us notice that by recording the transmitted signal with all the echoes and analysing individually the echoes (for thick samples) or the artificial periodic behaviour of the extracted parameter spectra (for thin samples), the knowledge of d is not required and even more can be deduced from the measurements [25].

8.5.3 Spectroscopy from Reflection Measurements

In the case of reflection spectroscopy, one employs the same principle as that in transmission and the complex reflection coefficient of the sample is obtained from two temporal measurements: $S(t)$ is obtained measuring the signal reflected on the

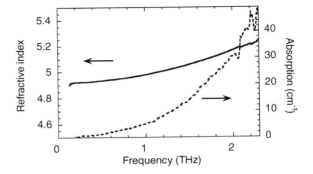

Fig. 8.11 The refractive index (*continuous line*) and absorption (*dashed line*) of a LaAlO$_3$ sample

sample, whereas $S_{ref}(t)$ is obtained using a perfect mirror for THz frequencies in the place of the sample. Usually, a flat metallic mirror is used to measure $S_{ref}(t)$. For most metals, especially noble metals, their huge permittivity makes the mirror reflectivity close to 100 %. In that case, the experimental difficulty is to place the sample at the very same position as the mirror to prevent any phase delay error, which can greatly corrupt the phase measurement of the reflection coefficient and then leads to wrong extracted optical parameters [26].

Principle

If one considers the same assumptions as described above on the sample and on the THz beam, the complex reflection coefficient of the sample under normal incidence writes:

$$R_{theo}(\omega) = re^{j\varphi_r} = \frac{\tilde{n} - 1}{\tilde{n} + 1} \frac{1 - e^{-2j\frac{\omega}{c}\tilde{n}d}}{1 - \left(\frac{\tilde{n}-1}{\tilde{n}+1}\right)^2 e^{-2j\frac{\omega}{c}\tilde{n}d}}. \tag{8.13}$$

In general, samples studied in a reflection scheme are either very thick or strongly absorbing. Thus, rebounds of the THz pulse inside the slab may be ignored, and the relation (8.13) simplifies into:

$$R_{theo}(\omega) = re^{j\varphi_r} = \frac{\tilde{n} - 1}{\tilde{n} + 1}. \tag{8.14}$$

Equation (8.14) can be analytically solved to obtain the expression of $\tilde{n} = n - j\kappa$ as a function of only the modulus r and the phase φ_r of the reflection coefficient:

$$n = \frac{1 - r^2}{1 - 2r\cos\varphi_r + r^2} \quad \text{and} \quad \kappa = \frac{2r\sin\varphi_r}{1 - 2r\cos\varphi_r + r^2}. \tag{8.15}$$

Figure 8.12 presents the refractive index (dots) and the absorption (open circles) of a *n*-type low-resistivity (0.038 Ω·cm) GaAs sample measured in reflection. Such a doped layer is not transparent enough to be characterised in transmission THz-TDS [27].

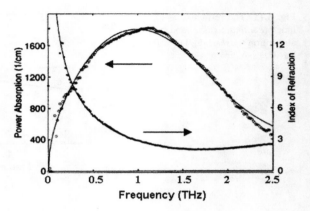

Fig. 8.12 The refractive index and absorption of a n-type 0.038 $\Omega \cdot$ cm resistivity GaAs sample together with a Drude-model fit (*continuous lines*) as adapted from [27]. Reprinted with permission from App. Phys. Lett. **72**, 3032 (1998). Copyright 1998 American Institute of Physics

8.6 Examples of Applications of THz-TDS

8.6.1 Sub-Wavelength ($\sim \lambda/100$) Films Characterization

As previously presented, classical THz-TDS transmission system is very useful to characterise samples with thickness d of the order or thicker than the wavelength $\lambda (\lambda = 300\,\mu m @ 1$ THz). For micron-thick films (a 1-μm film thickness is smaller than $\lambda/100$), the interaction length between the film and the THz field is weak. Consequently, the characterisation in the THz domain of such a film with a great precision remains a difficult challenge. For materials with a large dielectric constant such as ferroelectrics [28] or metals [29], the optical thickness ($d_{opt} = n \times d$) could be comparable to the THz wavelength even for nanometric films, and thus classical THz-TDS in transmission can be used to characterise them. However, d_{opt} is usually small [30] and classical THz-TDS in transmission is not sensitive enough to finely determine the material parameters. To solve this problem, smart methods have been proposed and demonstrated, based either on interferometry [31], on rapid sample dithering in the THz beam [32], or on locating the film in a metallic waveguide [33]. Amazing sensitivities have been reported since nanometer thick water films have been investigated [33]. Also, metallic plasmonic structures exhibiting super transmission are used to image thin films attached over the structures, which strongly affect the near-field pattern of the diffracted THz beam, and thus allows one to detect sub-micron thick layers [34, 35]. These solutions are based either on increasing measurements or TDS set-up sensitivity, or on electromagnetic-resonant devices that locally enhance the interaction between the THz field and the material.

Among resonant devices, planar dielectric waveguides, on which the film to be characterized is attached, are among the most sensitive ones [36, 37]. Such waveguides for the THz waves could simply consist of a transparent wafer. Because of the required waveguide thickness that should be of the order of the wavelength, the wafer is thick enough and thus mechanically robust to need no substrate, as it is

Fig. 8.13 A scheme of a grating device with the main geometrical parameters. Images are EBM pictures of the grating (**a**) and the section of a groove (**b**)

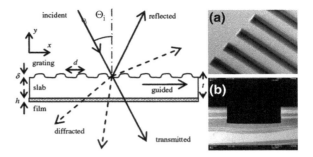

the case in the visible domain. However, phase matching should be realised between the incident THz beam and the guided THz wave. Among different solutions, the grating coupler, i.e. parallel grooves etched at the waveguide surface (Fig. 8.13), is the most versatile and compact one [38].

The theory of grating diffraction [39, 40] imposes the conservation of the component of the electromagnetic beam wave vector that is parallel to the waveguide surface:

$$k_i + p\frac{2\pi}{D} = k_p, \tag{8.16}$$

where p and D are, respectively the order of diffraction and the period of the grating, $k_i = \frac{2\pi}{\lambda}\sin\theta_i$ and $k_p = \frac{2\pi}{\lambda}\sin\theta_p$ are the parallel components of the respectively incident and diffracted wave vectors, while θ_i and θ_p are the angles of incidence and of diffraction. In a first approximation, the light is coupled into the waveguide when k_p is equal to the wave vector of the guided mode m, that is to say when:

$$k_i + p\frac{2\pi}{D} = \frac{2\pi}{\lambda}n_{\text{eff},m} \Leftrightarrow \sin\theta_i + p\frac{\lambda}{D} = n_{\text{eff},m}, \tag{8.17}$$

where $n_{\text{eff},m}$ is the effective index of mode m of the uncorrugated waveguide.

Under illumination at a fixed angle of incidence with a THz pulse that exhibits a broadband frequency spectrum, the device transmission spectrum shows Fabry-Pérot oscillations due to back and forth reflections in the slab, together with superimposed narrow absorption lines (the so-called m-lines) appearing at frequencies for which Eq. (8.15) is verified. These m-lines result from destructive interferences between the THz beam that is directly transmitted through the device and the guided beam that is progressively coupled out of the slab by the grating. When a thin film is deposited on the flat surface of the device (see Fig. 8.13), the guided modes are slightly modified, leading to a change in their wave vector; thus, one observes a variation in the spectrum transmitted by the covered device as compared to the bare device case.

At resonance, i.e. when a guided mode is excited, the interaction between the THz beam and the film is strongly enhanced, because the THz beam propagates in the device over a distance larger than the film thickness. The gain in sensitivity is proportional to the decay length of the guided mode in the grating region and to the overlap of the guided field and of the film, i.e. it depends on the guided mode number m. An accurate extraction of the thin-film parameters requires an almost noise-free experimental record of the transmitted THz signal, a rigorous electromagnetic code, and an actual device as perfect as possible. Here, the large dynamics of THz-TDS permits one to reduce the experimental noise to a very low level. To compute the diffraction efficiency, the differential method [39] that is known to nicely describe the electromagnetic response of dielectric gratings can be used. Figure 8.14 shows the transmission coefficient (amplitude) and the optical thickness ($d_{opt} = \omega/c\,\phi$, where ϕ is the phase of the transmission coefficient) of the bare device made of high-resistivity ($>4000\,\Omega \cdot$ cm) silicon. This waveguide is 285-μm thick and is illuminated under normal incidence. The complex permittivity of the silicon has been measured using a classical THz-TDS set-up and is almost constant over the whole frequency range ($\varepsilon_{Si} = 11.65 + j4.89 \times 10^{-5}$). The grating had been fabricated by chemical etching. The grooves are rectangular with period $D = 250\,\mu$m, groove depth $\delta = 41\,\mu$m and a filling factor 50 %, (see Fig. 8.13) leading to an optimised excitation of the guided modes in the range 0.1–1 THz. Let us notice that several m-lines, superimposed with the Fabry-Perot oscillations, can be observed in Fig. 8.14, each of them corresponding to the excitation of a guided mode through the different diffraction orders p. Continuous lines represent the response of the device calculated with the differential method, whereas dots stand for the experimental results.

To demonstrate the sensitivity of such a device, a Shipley-1828 photoresist layer deposited by spin coating on the flat slide of the device has been considered. The resist film thickness is $7.1 \pm 0.1\,\mu$m. The device with and without the film had been characterised using a classical THz-TDS shown in Fig. 8.7. The difference between the bare and covered grating results is very small and cannot be clearly observed on the whole spectrum figure. Thus the inset of Fig. 8.15 presents a zoom of the transmission spectrum around the m-line observed at 0.66 THz.

Moreover, to clearly emphasise the influence of the film, Fig. 8.15 shows the ratio of the amplitude of the transmission coefficients with and without the film. For most frequencies, the ratio is close to one, confirming that the influence of the film is weak.

Fig. 8.14 The transmission coefficient (amplitude) and optical thickness versus frequency of the bare grating device illuminated under normal incidence. Open circles correspond to measured data, while continuous lines are calculated

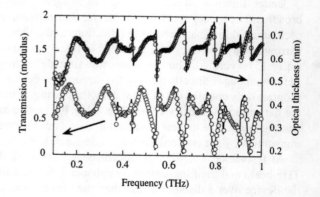

Fig. 8.15 The ratio of the spectra (amplitude) recorded with and without the film. The inset is a zoom of the transmission coefficient (amplitude) around 0.66 THz. Open circles and dashed line show, respectively, the measured data and the theoretical curve related to the bare device. Full circles and the continuous line are related to the device covered with the film

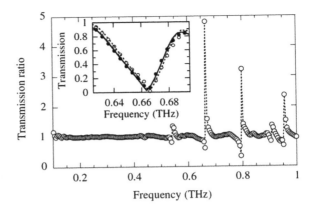

However, in the vicinity of the m-line frequencies, the ratio becomes larger than 1 and reaches almost 5 for the guided mode at 0.66 THz ($m = 3$, $p = +1$). For other m-lines, the ratio is smaller because the related guided modes ($m = 1, 2 \ldots$) do not overlap the film as efficiently as mode $m = 3$.

In order to extract the film parameters (n_{film} and α_{film}) around 0.66 THz, the experimental data recorded with the covered device were fitted using the differential method in which the only adjustable parameters are n_{film} and α_{film}. To solve this 2D-problem, we define the error function $A\left(|T_{\text{meas}}| - |T_{\text{theo}}|\right)^2 + (d_{\text{opt,meas}} - d_{\text{opt,theo}})^2$ and we integrate it over the m-line width (15 GHz). A is a weighting factor and the subscripts "meas" and "theo" mean, respectively, experimental and calculated. Figure 8.16 presents the error function value versus the real part ε' of the film permittivity for different values of the imaginary part ε''. A minimum is achieved for $\varepsilon' = 2.6$ and $\varepsilon'' = 0.4$, i.e. $n = 1.62$ and $\alpha = 34$ cm^{-1}. The precision is estimated to be $\Delta\varepsilon' = \Delta\varepsilon = 0.1$, leading to $\Delta n = 0.03$ and $\Delta\alpha = 9$ cm^{-1}. The material parameters deduced from a classical THz-TDS procedure using a 57-μm thick layer are $n = 1.59 \pm 0.02$ and $\alpha = 20 \pm 10$ cm^{-1} at 0.66 THz in agreement with the values obtained for the film. The precision of the method is determined by the frequency width of the m-line and the ability to measure it, which is in fact constrained by both experimental noise and frequency resolution. As the frequency resolution of THz-TDS is usually limited to a few GHz, one estimates that, using optimised grating groove shape that would lead to narrower m-lines, films five times thinner than the sample studied in this work could be characterised with the same precision.

8.6.2 Metamaterials in the THz Domain

Metamaterials are a subject of great interest because they can exhibit singular electromagnetic properties like a negative index of refraction [41]. In this paragraph, one is reporting on a periodic and metallic structure that shows a left-handed

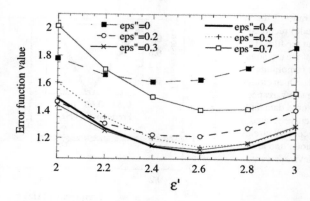

Fig. 8.16 The error function value versus the real part ε' of the film permittivity for different imaginary parts ε'' of the film permittivity. The lines are guides to the eyes

property and one is demonstrating that THz-TDS is well suited for characterising such metastructures. Here, the studied structures are made of a stack of 2D periodic elliptical and metallic holes arrays separated by dielectric layers (Fig. 8.17) [42].

The transmission and the reflection of the structure have been measured under normal incidence using a classical set-up of THz-TDS (Fig. 8.7). Figure 8.18 presents the propagation constant (real and imaginary part) determined from these THz-TDS measurements (magnitude and phase) for 3- and 5-layers devices. The electromagnetic behaviour depends on the number of metallic layers and exhibits a band gap around 500 GHz, which constitutes also the limit separating the left- and right-handed behaviours [43].

The left-handed area below 500 GHz is the signature of a negative type refractive index. Such behaviour opens the way to a new family of phenomena and applications. For example, one has demonstrated the principle of a negative refraction prism in the THz range [44]. The device is constituted of a stack of 10 metallic holes arrays layers according to a sequential mask shift to obtain a wedge-type prism shape (Fig. 8.19). The refraction of the prism has been measured using a "goniometric" THz-TDS

Fig. 8.17 The studied metamaterial structure

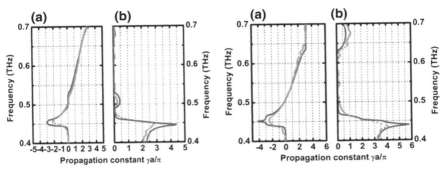

Fig. 8.18 The propagation constant (real part (**a**) and imaginary part (**b**)) of 3-(*left*) and 5-(*right*) layers devices. Measurements in *red*, simulations in *blue*. Reprinted with permission from J. App. Phys. 107, 074510 (2010). Copyright 2010 American Institute of Physics

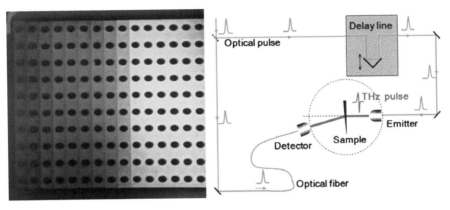

Fig. 8.19 A wedge-type prism (*left*) and the goniometric THz-TDS set-up (*right*)

set-up, in which a fibered detector, associated to a dispersion compensating system to prevent any enlargement of the femtosecond laser pulse in the fibre, can rotate around the studied sample.

Figure 8.20 (right) exhibits the measured transmission of the THz field versus the deflection angle at different frequencies. The refraction is characterised by a positive deflection angle to a negative one below 475 GHz, which corresponds to the transition from a right-handed to a left-handed behaviour as previously presented. This result is in agreement with the simulated refraction of the THz beam in this range of frequencies (Fig. 8.20-left), which indicates a negative refraction around 490 GHz.

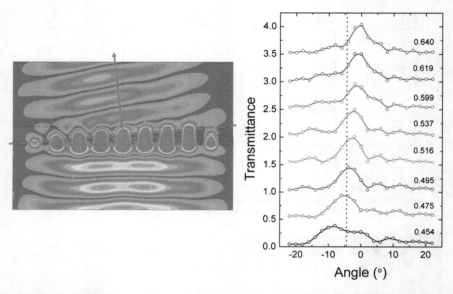

Fig. 8.20 Negative refraction of the prism simulated around 490 GHz (*left*) and measurements of the refraction of the prism for different frequencies (*right*)

8.7 Uncertainties and Precision of THz-TDS

The precision on the optical constants determination by mean of THz-TDS is limited by several error sources related to the extraction procedure [45, 46]. First, this procedure, as explained in the previous section, is based on the hypothesis of a flat and homogeneous sample with parallel faces, which is illuminated over normal incidence by a parallel THz beam. Any deviation from this hypothesis results in a systematic error [47]. Second, the THz signal is affected by noise that leads to an additional imprecision on $\tilde{n}$. Let us first treat the systematic errors. To check the influence of the incidence angle, Eqs. (8.10) and (8.11) are written for an oblique incidence, and then are derived versus the angle of incidence. This leads to very long equations that are not reproduce here, but which can be summarised as follows: For a typical 1-mm thick sample around 1 THz, an angular tilt of 1 degree as compared to normal incidence induces a change of the order of 0.01% for either the transmission or reflection coefficient. The influence of the sample thickness d is rather strong. Indeed, a small variation Δd of the thickness d changes the transmission and reflection coefficients as:

$$\frac{\Delta T}{T} \approx \frac{\omega}{c} |\tilde{n}| \, \Delta d, \quad \frac{\Delta R}{R} \approx 8 \frac{\omega}{c} \left| e^{j2\frac{\omega}{c}\tilde{n}d} \frac{(\tilde{n}-1)^2}{(\tilde{n}+1)^2} \right| \Delta d \qquad (8.18)$$

At 1 THz and for common materials ($\kappa < 1$, i.e. $\alpha < 400\,\text{cm}^{-1}$ at 1 THz), these relations simply write:

$$\frac{\Delta T}{T}\,(\%) \approx 2n\,\Delta d_{\mu m}, \quad \frac{\Delta R}{R}\,(\%) \approx 0.3\left(\frac{n-1}{n+1}\right)^2 \Delta d_{\mu m}, \tag{8.19}$$

where $\Delta d_{\mu m}$ is Δd given in μm. Thus, an error of $\Delta d_{\mu m} = 10\,\mu$m leads to a wrong estimation of the transmission by about some tens of percents, while it is only a few percents in reflection, as most of the reflected signal corresponds to the reflection at the first sample face, and thus does not depend on the sample thickness. The effect of the THz beam shape is more difficult to evaluate. Nevertheless, a convergent beam can be described as a sum of plane waves impinging the sample under an angle. The order of magnitude of this effect is thus related to the one of a tilted sample, i.e. it is smaller than 1 % for a divergence of a few degrees.

Concerning the random errors originating from noise, let us treat here the case of a set-up with PSW antennae, from which the case of electro-optic detection can be easily extrapolated. As already described in the section devoted to the SNR of the THz-TDS set-up, the noise sources are the laser intensity that may fluctuate, the shot noise of the THz beam at the receiver and the noise in the electronics (Eq. (8.9)). Let us consider the measure of a complex signal S, of amplitude s and phase ϕ, either in reflection or in transmission. Because of noise, the measured signal S_{meas} is equal to:

$$S_{\text{meas}} = S + S_N \Leftrightarrow s_{\text{meas}}e^{j\phi_{\text{meas}}} = se^{j\phi} + s_N e^{j\phi_N} s_N << s, \tag{8.20}$$

where the subscript N is for noise. It is easy to show that:

$$s_{\text{meas}} = s + s_N \cos(\phi_N - \phi), \quad \phi_{\text{meas}} = \text{atan}\left(\frac{s\sin\phi + s_N \sin\phi_N}{s\cos\phi + s_N \cos\phi_N}\right). \tag{8.21}$$

From these relations, one deduces the standard deviations σ_s and σ_ϕ of s and ϕ, which are equal to the errors Δs and $\Delta\phi$:

$$\Delta s \equiv \sigma_s^2 = \overline{(s_{\text{meas}} - s)^2} = \overline{s_N^2 \cos^2(\phi_N - \phi)} = \overline{s_N^2}/2, \quad \Delta\phi \equiv \sigma_\phi^2 = \sigma_s^2/s^2, \tag{8.22}$$

where the average value of x is noted $\bar{x}$.

On the other hand, by derivation of Eqs. (8.10) and (8.15) in the case where one can time window the first transmitted or reflected pulse, we obtain analytical expression for the uncertainties over n and κ that depends on the standard deviation of the transmission or reflection coefficients:

$$\Delta n = \Delta\kappa = \frac{(2(n+1+\kappa)+((n+1)^2+\kappa^2)\beta)}{4(1+\kappa\beta)+((n+1)^2+\kappa^2)\beta^2}\frac{(n+1)^2+\kappa^2}{4}e^{\kappa\beta}\sigma_t,$$

$$\Delta n = \Delta\kappa = \frac{\left(|\chi^2-1-\sqrt{\left((\chi^2-1)^2+4\kappa^2\right)\left((\chi+1)^2-4n^2\right)}|+4n\kappa\right)}{\left(\sqrt{((\chi-1)^2+4\kappa^2)(\chi+1)^3}-(\chi-1)\sqrt{\chi-2n+1}\right)^2}\sigma_r, \tag{8.23}$$

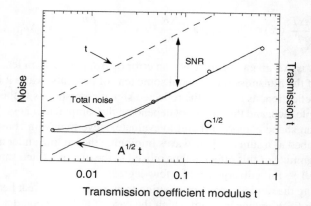

Fig. 8.21 Different contributions to noise versus the transmission of the silicon sample under test as given by Eq. (8.25). The *open circles* are experimental data of the standard deviation of measured amplitude at 250 GHz. The modulus is also given as a guided to the eye

with $\beta = d\omega/c$ and $\chi = n^2 + \kappa^2$. The standard deviation of the transmission or reflection coefficients is calculated from the experimental data. In the case of transmission, let S_i and S_T be the incident and transmitted signals, respectively. One gets:

$$T_{\exp} = \frac{S_T + S_{NT}}{S_i + S_{Ni}} \approx T + \frac{S_{NT} - T S_{Ni}}{S_i} \rightarrow \sigma_t = \frac{\sqrt{\sigma_{S_T}^2 + t^2 \sigma_{S_i}^2}}{s_i} \qquad (8.24)$$

A similar expression is obtained in reflection. Thus, through several measurements of the signals, the standard deviations are obtained using Eq. (8.22), and thus Δn and $\Delta \kappa$ are determined through Eq. (8.21). As we are dealing with THz-TDS, the available spectrum is limited to several THz, and the signal magnitude decreases strongly at lower and higher frequencies, leading to a strong noise at the borders of the spectrum. A deeper insight in the physical origin of the noise allows us to get the dependence of the noise on the signal strength.

By writing the total noise with and without sample, one obtains a quadratic dependence of the noise with the transmission (reflection) coefficient:

$$\sigma_{S_x}^2 = x^2 \sigma_E^2 + \sigma_{shot}^2 + \sigma_D^2 = x^2 \sigma_E^2 + 2q s_x \Delta f + \sigma_D^2$$
$$\sigma_{S_i}^2 = \sigma_E^2 + 2q s_i \Delta f + \sigma_D^2$$
$$\Rightarrow \sigma_x^2 = A(\omega) x^2 + B(\omega) x + C(\omega) \quad x = r \text{ or } t \qquad (8.25)$$

The coefficients A, B and C are characteristic of each THz-TDS set-up. In Fig. 8.21, one plots the experimental noise measured for different values of $t(\omega)$ in a transmission experiment. Each point has been obtained by calculating the standard deviation of a dozen of records of the THz amplitude at 250 GHz. The experimental data are

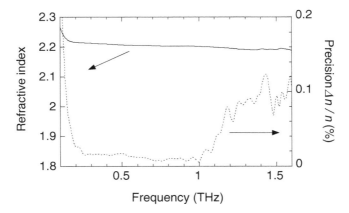

Fig. 8.22 Refractive index versus frequency of a slab of solid glue (Stycast®) and uncertainty due to noise

perfectly fitted by Eq. (8.23). One can notice that for the lower values of t ($t < 0.02$), SNR decreases with $t(\omega)$ and for moderate to highly transparent samples ($t > 0.02$), SNR remains constant whatever the transmission of the sample is. This latter situation appears to be the best one in terms of measurement but, on the other side, a TDS experiment is basically a differential method as explained in the first section. Thus, the accuracy on the results varies as the difference between $S(\omega)$ and $S_{\text{ref}}(\omega)$. In other words, it is preferable to deal with rather low values of t. Taking into account, this consideration together with the dependency of the SNR on t, one can conclude that the optimum sample transmission is the one for which SNR starts to decrease because of noise. In Fig. 8.21, it corresponds to a value of t of about 0.03 (at 250 GHz for the set-up employed in this work).

The validity of our complete model of uncertainty has been demonstrated by comparison with the experiment, in transmission [45] and in reflection [48]. Figure 8.22 shows the spectrum of the refractive index of a 1.62-mm thick slab of solid glue (Stycast®) as well as the uncertainty due to noise, measured in transmission. The uncertainty due to noise is ~0.1 %. In this case, the main source of error is the uncertainty on the sample thickness with value $\Delta d = 20\,\mu$m, which gives $\Delta n \sim 1$ %.

Therefore, THz-TDS is a very precise method to determine the refractive index of a sample. The precision on n is typically smaller than 1 %, depending on the set-up characteristics, but also on the optical thickness d_{opt} of the sample. Indeed, if the sample is too thin, the induced phase change is small and the uncertainty on n increase. On the other hand, the precision on the absorption coefficient depends strongly on the detected signal amplitude, which may be rather weak in transmission for highly absorbing media. Thus one may guess that for these materials, reflection measurements are preferable while for high transparent samples, transmission will give more accurate results. This is theoretically checked [48] by comparing the values of Δn and $\Delta \kappa$ calculated in transmission and reflection with Eq. (8.23) together

Fig. 8.23 Contour map, versus n and κ, of the limit when both reflection and transmission THz-TDS lead to the same precision on Δn. Each curve is plotted for a different value of the parameter $\beta = d\omega/c$. Above each curve, reflection is more precise

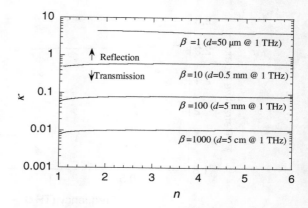

with the noise expression (8.25). One has plotted on Fig. 8.23 the contour map, versus n and κ, of the limit when both techniques lead to the same precision (here Δn). Above the curve, reflection is more precise. As transmission is function of the sample thickness, several plots have been drawn for different values of $\beta = d\omega/c$. The reflection technique is definitively better than the transmission one for thick and absorbing samples.

8.8 Comparison with FTIR Measurements

Even if THz-TDS has became in the last 10 years a very useful technique to investigate the THz range, Fourier Transform Spectroscopy in the infrared (FTIR) remains a mature and very popular technique, which employs reliable and easy-to-use apparatuses. It is interesting to compare these two techniques. In terms of bandwidth, FTIR is of course better than THz-TDS as it can be used from the visible to the far infrared domain, but often to the detriment of its sensitivity. In general, THz-TDS has a better SNR than FTIR below 3 THz ($100\,\text{cm}^{-1}$), whereas FTIR is preferable over 5 THz [49]. Thanks to the progress of THz-TDS set-ups, this limit may shift to higher frequencies in a near future. For example, first results of TDS with generation and detection of THz waves, up to 20 THz, by optical breaking in air seem very promising [50, 51].

The advantage of THz-TDS is the possibility of measuring not only the intensity of the THz signal but its magnitude and its phase, allowing determining not only the absorbance of a sample, but its complex refractive index. This could be also achieved in FTIR, by putting the sample in one arm of the interferometer (differential FTIR). Nevertheless, the positioning of the sample as well as the stability of the source and the SNR are tricky points that have made this technique not so common.

In THz-TDS, the possibility of time windowing opens the way to new applications like time-of-flight and time-resolved measurements. Recently, the generation

Table 8.1 Comparative performances between THz-TDS and FTIR techniques

	THz-TDS	FTIR
DR of power ($f < 3\,THz$)	up to 10^8	~300
Bandwidth	0.1–3 THz (regular TDS)	Far to mid IR
Frequency resolution	~0.1 cm^{-1}	~0.1 cm^{-1}
Source stability	Good	Not good
Acquisition time	Minutes or less	Minutes
Particularity	Time-gated Coherent pulse High SNR	Broad spectrum
Availability	Initial stage	Mature, easy

of powerful THz pulses with amplified femtosecond lasers has allowed to start the study of non-linear effects in the THz range [52], which is definitely not possible with classical FTIR.

Table 8.1 summarises the advantages and drawbacks of the two techniques [49, 50].

8.9 Conclusion

THz-TDS is really a wonderful tool to study and characterise materials and devices in the far infrared. From an academic point-of-view, it is a perfect illustration of the application of the Fourier transform. In this sense, the technique is very close to the popular Fourier Transform (FTIR) spectroscopy. In THz-TDS, performing the temporal sampling is equivalent to deal with an interferometer in the frequency domain. In fact, the classical THz-TDS set-up in transmission shown in Fig. 8.7 resembles a Mach-Zehnder interferometer in one arm of which the laser light is transformed into a THz beam.

From the practical point of view, the most striking performances of THz-TDS are its large bandwidth associated to unrivalled SNR and dynamics. These features are inherently linked to the use of very short and very stable optical pulses combs delivered by modern mode-locked lasers. Indeed, the femtosecond optical pulses permit to generate very high THz frequencies and to reject all the background noise by switching on the receiver only when necessary, i.e. when the THz signal impinges the detector. This temporal approach may also be used for THz time-resolved study, in which a third laser beam excites the sample synchronously with the THz pulses [53, 54]. Together with the huge amount of applications of the THz waves, these excellent performances explain why commercial THz-TDS apparatuses are now available. The variety of such commercial equipments is rather wide, from laboratory set-ups to movable compact systems that are power supplied by battery.

References

1. E.D. Palik, J. Opt. Soc. Am. **67**, 857 (1977)
2. Darrel T. Emerson, IEEE Trans. Microw. Theory Tech. **45**, 2267 (1997)
3. R.J. Bell, *Introductory Fourier Transform Spectroscopy* (Academic Press, New York, 1972)
4. K.H. Yang, P.L. Richards, T.R. Shen, Appl. Phys. Lett. **19**, 320 (1971)
5. D.H. Auston, Appl. Phys. Lett. **26**, 101 (1975)
6. G. Mourou, C.V. Stancampiano, A. Antonetti, A. Orszag, Appl. Phys. Lett. **39**, 295 (1981)
7. D.H. Auston, K.P. Cheung, P.R. Smith, Appl. Phys. Lett. **45**, 284 (1984)
8. Y. Pastor, G. Arjavalingam, J.M. Halbout, G.V. Kopcsay, Elec. Lett. **25**, 523 (1989)
9. M. van Exter, C. Fattinger, D. Grischkowsky, Opt. Lett. **14**, 1128 (1989)
10. S.L. Dexheimer, *Terahertz Spectroscopy: Principles and Applications* (CRC Press, New York, 2007)
11. J. Takayanagi et al., Opt. Express **17**, 7533 (2009)
12. A. Sell, A. Leitenstorfer, R. Huber, Opt. Lett. **23**, 2767 (2008)
13. Q.W. Wu, X.-C. Zhang, Appl. Phys. Lett. **71**, 1285 (1997)
14. A. Krotkus, K. Bertulis, M. Kaminska, K. Korona, A. Wolos, J. Siegert, S. Marcinkevicius, J.-F. Roux, J.-L. Coutaz, IEE Proc. Optoelectron **149**, 111 (2002)
15. L. Duvillaret, F. Garet, J.-F. Roux, J.- L. Coutaz, IEEE J. Sel. Top. Quant. Electron. **7**, 615 (2001).
16. J.-F. Roux, J.-M. Delord, J.-L. Coutaz, Phys. Status Solidi **C6**, 2843 (2009)
17. E. Castro-Camus, L. Fu, J. Lloyd-Hughes, H.H. Tan, C. Jagadish, M.B. Johnston, J. Appl. Phys. **104**, 053113 (2008)
18. E. Matsubaraa, M. Bitoh, H. Shimosato, M. Ashida: Ultrabroadband electric field generation and detection from far infrared to optical communication frequency. Paper presented at the 35th international conference in infrared millimeter and teraHertz waves (IRMMW-THz 2010), Roma, 5–10 Sept 2010
19. G. Klatt, R. Gebs, H. Schäfer, M. Nagel, C. Janke, A. Bartels, T. Dekorsy, IEEE J. Sel. Top. Quant. Electron. **17**, 158 (2011)
20. F. Blanchard, A. Doi, T. Tanaka, H. Hirori, H. Tanaka, Y. Kadoya, K. Tanaka, Opt. Express **19**, 8277 (2011)
21. Let us note that, among all the contributions of white noise, the thermal noise in the PSW, which is depending on the detector electrical resistance, should in principle vary with the optical power as demonstrated in [17, 22]
22. E. Castro-Camus, J. Lloyd-Hughes, L. Fu, H.H. Tan, C. Jagadish, M.B. Johnston, Opt. Express **15**, 7047 (2007)
23. L. Duvillaret, F. Garet, J.-L. Coutaz, IEEE J. Sel. Top. Quant. Electron. **2**, 739 (1996)
24. I. Pupeza, R. Wilk, M. Koch, Opt. Express **15**, 4335 (2007)
25. L. Duvillaret, F. Garet, J.-L. Coutaz, Appl. Opt. **38**, 409 (1999)
26. A. Pashkin, M. Kempa, H. Nemec, F. Kadlec, P. Kuzel, Rev. Sci. Instrum. **74**, 4711 (2003)
27. T.I. Jeon, D. Grischkowsky, Appl. Phys. Lett. **72**, 3032 (1998)
28. M. Misra, K. Kotani, T. Kiwa, I. Kawayama, H. Murakami, M. Tonouchi, Appl. Surf. Sci. **237**, 421 (2004)
29. F. Garet, L. Duvillaret, J.-L. Coutaz, THz time-domain spectroscopy of nanometric-thick gold layers. Paper presented at the 29th international conference in infrared millimeter and teraHertz waves (IRMMW-THz 2004), Karlsruhe, Sept 2004
30. S. Labbé-Lavigne, S. Barret, F. Garet, L. Duvillaret, J.-L. Coutaz, J. Appl. Phys. **83**, 6007 (1998)
31. S. Krishnamurthy, M.T. Reiten, S.A. Harmon, R.A. Cheville, Appl. Phys. Lett. **79**, 875 (2001)
32. Z.P. Jiang, M. Li, X.C. Zhang, Appl. Phys. Lett. **76**, 3221 (2000)
33. J. Zhang, D. Grischkowsky, Opt. Lett. **29**, 1617 (2004)
34. M. Tanaka, F. Miyamaru, M. Hangyo, M. Hangyo, T. Tanaka, M. Akazawa, E. Sano, Opt. Lett. **30**, 1210 (2005)

35. F. Miyamaru, S. Hayashi, C. Otani, K. Kawase, Y. Ogawa, H. Yoshida, E. Kato, Opt. Lett. **31**, 1118 (2006)
36. F. Garet, Y. Laamiri, J.-L. Coutaz, Frequenz **62**, 123 (2008)
37. Y. Laamiri, F. Garet, J.-L. Coutaz, Appl. Phys. Lett. **94**, 071106 (2009)
38. J.-F. Roux, F. Aquistapace, F. Garet, L. Duvillaret, J.-L. Coutaz, Appl. Opt. **41**, 6507 (2002)
39. P. Vincent, in *Electromagnetic Theory of Gratings*, ed. by R. Petit. Topics, in Current Physics, vol. 22 (Springer-Verlag, Berlin, 1980), pp. 101–121
40. M. Nevière, E. Popov, R. Reinisch, G. Vitrant, *Electromagnetic Resonances in Nonlinear Optics* (Gordon and Breach Science Publishers, Amsterdam, 2000)
41. S. Anantha Ramakrishna, T. Grzegorczyk, *Physics and application of negative refractive index materials* (SPIE Press and CRC Press, New York, 2009)
42. C. Croënne, F. Garet, E. Lheurette, J.-L. Coutaz, D. Lippens, Appl. Phys. Lett. **94**, 0133112 (2009)
43. S. Wang, F. Garet, K. Blary, C. Croënne, E. Lheurette, J.-L. Coutaz, D. Lippens, J. Appl. Phys. **107**, 074510 (2010)
44. S. Wang, F. Garet, K. Blary, C. Coënne, E. Lheurette, J.-L. Coutaz, D. Lippens, Appl. Phys. Lett. **97**, 181902 (2010)
45. L. Duvillaret, F. Garet, J.-L. Coutaz, J. Opt. Soc. Am. **B17**, 452 (2000)
46. M. van Exter, D. Grischkowsky, IEEE Trans. Microw. Theory Tech. **38**, 1684 (1990)
47. W. Withayachumnankul, B.M. Fischer, H. Lin, D. Abbott, J. Opt. Soc. Am. **B25**, 1059 (2008)
48. F. Garet, M. Bernier, B. Blampey, J.-L. Coutaz, Accurate characterization of materials by transmission and reflection THz time-domain spectroscopy. Paper presented at the 1st conference teranano 2011, University of Osaka, Japan, 24–27 Nov 2011
49. P.Y. Han, M. Tani, M. Usami, S. Kono, R. Kersting, X.-C. Zhang, J. Appl. Phys. **89**, 2357 (2001)
50. I.-C. Ho, Xiaoyu Guo, X.-C. Zhang, Opt. Express **18**, 2872 (2010)
51. M. Zalkovskij, R. Malureanu, A. V. Lavrinenko, P. U. Jepsen, D. Savastru, A. Popescu, Ultrabroadband THz spectroscopy of disordered materials. Paper presented at the 1st conference teranano 2011, University of Osaka, Japan, 24–27 Nov 2011
52. P. Gaal, K. Reimann, M. Woerner, T. Elsaesser, R. Hey, K.H. Ploog, Phys. Rev. Lett. **96**, 187402 (2006)
53. For this chapter to remain short, we do not describe this technique. The interested reader can look at: H. Němec, F. Kadlec, S. Surendran, P. Kužel, P. Jungwirth, J. Chem. Phys. **122**(10), 4503 (2005)
54. H. Němec, F. Kadlec, C. Kadlec, P. Kužel, P. Jungwirth, J. Chem. Phys. **122**(10), 4504 (2005)

Chapter 9
Terahertz Frequency Security Systems and Terahertz Safety Considerations

R. Appleby and J. M. Chamberlain

Abstract The use of THz technology for security applications is reviewed and recent developments discussed. It now seems unlikely that THz spectroscopy can be used to conduct identification of materials at standoff ranges. THz technology is becoming more widely deployed in other applications including non-destructive testing, medicine, food and drink analysis and communication and it is now becoming apparent that safety standards are required to limit the exposure to THz radiation. From a review of the current literature, it is clear that, under certain irradiation conditions, there may be mechanisms operating whereby THz radiation can influence the genetic process.

Keywords Terahertz safety · Terahertz security · Exposure limits · Irradiation · Standoff

9.1 Overview

The use of terahertz (THz) frequency radiation in security systems stems from two properties: the first is the ability to penetrate dielectrics such as clothing; and the second is that many materials have spectroscopic signatures. The THz region, which here we will take as 0.1 to 10 THz, therefore not only offers the ability to detect threats under clothing but also the possibility of identifying them. It also has other applications as relatively small amounts of gaseous, liquid or solid material can be detected and identified from their spectra.

R. Appleby
Queen's University Belfast, Belfast, UK

J. M. Chamberlain (✉)
University of Durham, Durham, UK
e-mail: martyn.chamberlain@durham.ac.uk

M. Perenzoni and D. J. Paul (eds.), *Physics and Applications of Terahertz Radiation*,   233
Springer Series in Optical Sciences 173, DOI: 10.1007/978-94-007-3837-9_9,
© Springer Science+Business Media Dordrecht 2014

THz technology is now becoming more widely used in everyday applications such as non-destructive testing [1], medicine [2], food and drink analysis [3] and communications [4]. Given the potential economic value of the THz market place [5], increasing importance attaches to determining the safe limits of exposure to THz radiation; and that the potential risks of exceeding such limits are well-established in circumstances where experimental data are often conflicting.

In this chapter, we will review the security threats where THz imaging can have an impact, summarise the physical principles which underpin this capability, comment on available commercial systems and outline new developments where they are applicable to security systems. Safety aspects of all THz systems are also reviewed.

9.2 Nature of Security Threats, Ethical Considerations

The threats faced in modern life from asymmetric warfare vary considerably in different scenarios from people-borne, buried, roadside and vehicle-borne devices. THz systems are not suitable for addressing all threats, as there are some materials that this radiation cannot penetrate such as metal and water. The following list highlights those situations where THz systems could provide a useful function:

- Detection of weapons and contraband under clothing;
- Detection of small amounts of energetic material in containers;
- Detection of hazardous chemical and biological vapours;
- Detection of illegal immigrants in soft-sided vehicles;
- Perimeter monitoring;
- Detection of buried objects due to changes in soil moisture content.

Perhaps the best known of all these applications is the use of body scanners in aviation security. Since September 11th 2001 ("9/11"), the USA has installed 488 systems in 78 airports [6], of which some use X-ray and some millimetre wave (MMW) detection (30–300 GHz). The worldwide deployment is probably three times this number. These systems bring with them ethical and privacy issues as some of them produce graphic images. The ethics associated with body scanners and their impact on system design and operation have been reported [7]. To overcome some of these concerns, various manufacturers have introduced automatic target recognition software which does not display the MMW image, but either displays the objects on a video image or on an avatar.

9.3 Physical Principles

The physical principles of operation of MMW and THz systems have been extensively reviewed elsewhere [8–10] and so only the briefest summary will now be given. At MMW frequencies, gases will exhibit rotational spectra; the spectra of solids in the MMW region, unlike gases, are essentially flat. Therefore, contrast in MMW images does not rely directly on the presence of discrete absorption bands; rather it is obtained from the different reflectivity exhibited by various materials in the scene illuminated by black body radiation from the cold sky. (Note that the sky temperature at zenith can be as low as 100 K in good conditions). Images at THz frequencies, on the other hand, derive their contrast from the specific vibrational, rotational or librational modes of molecules or groups of molecules for solids, liquids and gases. The fundamental distinction between active and passive imaging is well known [10]; in practice the low level of background illumination from the sky will preclude passive imaging above about 250 GHz. Another distinction between the two regimes is that, at THz frequencies, there are significant practical advantages in terms of higher resolution and smaller effective antenna size. A crucial consideration in system design is the attenuation of radiation due to absorption and scattering in the atmosphere. At MMW frequencies, specific molecular absorption leads to "windows" between bands (e.g. at 35, 94 and 140 GHz); scattering is largely due to raindrops, but smoke and fog particles have little effect. At THz frequencies, molecular absorption is even greater giving rise to very few windows and reaches a level of around $300\,\mathrm{dBkm}^{-1}$; this value effectively prevents THz security systems operating at standoff distances greater than ten metres or so under normal circumstances.

9.4 Review of Previous Topical Reviews (2006–2010)

MMW and THz technology have been applied at short range in the form of portals or at standoff ($>1\,\mathrm{m}$). The optical properties of materials, atmospheric propagation and the techniques of standoff detection of weapons and contraband were comprehensively reviewed by Appleby and Wallace [8]. The application of this technology to biological and chemical sensing was also reported by Woolard [11].

The discovery of the first spectral features in the THz region in a number of common materials and explosives were first reported in 2003 by Kemp [12]. The potential of this technique was subsequently reviewed as more information emerged and the technology started to mature [13–16]. In a recent review, also by Kemp [17], the use of spectroscopy for countering people-borne energetic devices was seen as "a bridge to far": that is, in any practical scenario, it is unlikely to provide genuinely useful information. This is a result of several factors which compound together and include:

- Energetic materials can only be observed in reflection;
- Scattering which can result from the rough surface of the material or clothing dissipates the signal;
- Atmospheric absorption which can be severe in humid environments.

It now seems unlikely that systems based solely on spectroscopy will ever transfer from the laboratory to practical arrangements used against people-borne devices. However, despite this view, it is noted that Kong et al. and Dimelik et al. [18, 19]

have demonstrated that small (mg) amounts of energetic material can be detected with THz spectroscopy. This quantity is much greater than can be detected by other techniques; THz methods might thus still be useful for detecting small amounts of material which contaminate the location of an explosive device and may not be accessible by other methods.

The lower frequency range (0.3−1 THz) of the THz band is sometimes referred to as the submillimetre wave (SMMW) region; SMMW radiation will penetrate clothing and operate in humid locations. Compared to systems operating above 1 THz SMMW systems will have a larger physical size but will still typically have reduced spatial resolution due to the lower operating frequency. This disadvantage may be offset in some cases by using synthetic apertures [20]. It is not possible to unambiguously identify materials in the SMMW range, as there are no clear spectral features [21]; however shape, which can be discerned using SMMW radiation, gives a strong indication of the need to investigate further a suspicious object under test. The development over the last ten years of SMMW systems for people screening, the associated phenomenology and the enabling technology were also reported as a chapter in another recent publication in this Series [9] that also reviews THz spectroscopy.

9.5 Review of Available Commercial and Near-to-Market Systems

Very few systems have been deployed recently, but it is worth noting the work by Rhode and Schwartz on 77 GHz phased arrays and Smiths Detection who have developed a system based on a reflect array.

9.5.1 Phased Array Radar 77 GHz

Rhode and Schwartz made use of a low cost chipset based on SiGe [22] developed for car collision avoidance to develop a phased array radar that can be deployed as a security scanning portal [23]. This system consists of several flat panels which can be arranged to give a full view of the person being screened.

9.5.2 Reflect Arrays

These devices are in essence a programmable mirror consisting of an array of small antennas and associated circuits. The direction of the reflected beam is then determined by a set of time delays under digital control. Coherent radiation is typically used and so the time delays can be replaced by a set of phase shifters. Agilent first reported the application of this technology to security scanning [24] and it was subsequently developed by Smiths [25].

9.6 New Developments

In the past 5 years several new techniques and technologies have emerged which are relevant to security systems and are summarised below.

9.6.1 High Frequency Radar

Frequency Modulated Continuous Wave (FMCW) radars at 600 [26, 27] and 300 GHz [28] have been reported with the 600 GHz system detecting a gun under clothing at 25 m. If the image is based solely on the target's reflectivity the discrimination is poor due to speckle and glint; but, by exploiting the bandwidth available at these frequencies, this allows subcentimetre range gating to be used to separate items from the body based only on their range. The challenge here is to develop multi-receiver architectures to enable real time imaging.

9.6.2 Quantum Cascade Lasers (QCLs)

QCLs have become more than an "idle curiosity"; the current state of art is reviewed elsewhere in this Volume, but CW outputs greater than 100 mW at 4–5 THz and a few mW at 1–2 THz are now achievable. These devices have been used as illuminators to produce images when the scene is viewed by a thermal imaging camera which, although being used out of band, still has sufficient sensitivity to produce an image. Some of these cameras have been modified to enhance their performance in the THz region [29–31]. Electronic tuning has also been achieved giving rise to laboratory demonstrations of spectroscopy [32, 33].

It is also possible to use a QCL as part of radar system, providing both transmitter power and also the local oscillator to pump a mixer. The short length of these devices does produce multimode output which needs to be converted into a single mode to couple with the mixer correctly. This has been achieved by Hubers et al. [34] who used a 2.4 THz device to pump a hot electron bolometer mixer.

9.6.3 Cooled Bolometers

Radio astronomy has been making use of devices cooled to liquid helium temperatures and below to deliver some of the best detectors available in this frequency band. Some of these technologies are starting to be applied to security imaging. Luukaneen with colleagues at NIST [35–37] have reported the use of linear arrays of Nb and NbN up to 64 elements long which have been used as a basis for real time passive imaging systems. This system operates at 4 K and uses a pulsed tube cooler and has a bandwidth from 0.2 to 0.8 THz. Grossman [38] combined this camera with a frequency selective surface to realise a multispectral imager. Heinz [39, 40] has produced world leading performance from 20 superconducting transition-edge sensors (TES) operating at 450–650 mK and has achieved a thermal sensitivity of 0.1–0.4 K operating at 350 GHz at 10 frames per second.

9.6.4 Photonic Generation

THz radiation has been generated by beating together infrared radiation from a solid-state laser developed for high speed telecommunications [41]. Phase modulation was limited to 110 kHz; this is three times faster than a mechanical system and, when implemented with an interferometric imaging system, has the potential for video rate imaging. A spectrometer has also been developed based on photomixing [42] that operates in the 0.1–1.2 THz region and is capable of resolving the spectra of gases.

9.6.5 Air Breakdown Coherent Detection

Early work at Rensselaer Polytechnic [43, 44] and at Frankfurt University [44] identified that THz radiation can be generated in a plasma formed by focusing two infrared lasers at slightly different frequencies at a distant point. This was given the acronym Air Breakdown Coherent Detection (ABCD) by the Rensselaer group. As it is the infrared radiation that propagates in air, rather than a THz beam, atmospheric losses are overcome. Detection is achieved [45] by coherently manipulating the fluorescence emitted from asymmetrically ionised gases. The disadvantage of this technique is that the lasers used are not eye-safe and therefore could not be utilised to image people. However, given that the THz beam can be generated orthogonally to the infrared beam, this approach might be deployed for scanning horizontal surfaces for contamination by energetic materials or other chemicals.

9.6.6 CMOS

Direct and heterodyne detection has been reported operating at 0.65 THz using arrays of CMOS detectors [46]. Schuster et al. [47] also report array fabrication in a low-cost 130 nm silicon CMOS technology. Using an integrated bow-tie antenna, these authors achieve a record responsivity above $5\,kVW^{-1}$ and a noise equivalent power below $10\,pWHz^{-0.5}$ in the atmospheric window around 300 GHz and at room temperature. The use of this technology also appears possible up to higher frequencies and should be considered as the main contender for future low cost imaging systems in the SMMW region.

9.6.7 Silicon Germanium

The first receiver based on this technology with a performance suitable for imaging was reported at 0.1 THz [48]. A thermal sensitivity of 0.3–0.4 K was reported with a 30 ms integration time and a $\sim$20 GHz bandwidth. This technology was used to develop a chip set for collision avoidance radar which was subsequently developed into a security screening portal based on a phased array (see Sect. 9.5.1).

9.7 Terahertz Safety

9.7.1 Introduction

The question of THz system safety will now be considered. This is especially timely as recent theoretical modelling and experimental work [49, 50] has indicated that THz radiation is capable of "unzipping" the double strands of DNA, thereby interfering with the transcription process and, perhaps, leading to unwanted and potentially dangerous genetic changes. Experiments are poorly evidenced; and it is worth quoting again [51] a comment, originally made in 1996 [52], that "…the non-reproducibility of results and the non-robustness of effects…seem especially vexatious…there is no repeatable pattern…". Finally, it should be remarked that great importance also attaches to the development of more credible models of both interaction and energy dissipation mechanisms, especially as there has been much debate concerning the role of subtle coherent processes [53] which are claimed to have far-reaching implications [54].

The THz energy scale corresponds to the energy available at room temperature, i.e. to several milli-electron volts. This is typical of the excitation energies for many biological processes at the lowest levels of organisation and so electromagnetic radiation (EMR)/ biological material interactions seem very likely. Such processes include: proton tunnelling for enzyme active sites [55], the collective motion of DNA base pairs along the hydrogen-bonded backbone [56]; protein conformation [57]; the mechanical vibration of cell membranes [53, 58]; and the nonlinear thermal fluctuations that are thought to initiate DNA transcription [59]. Moreover, THz EMR can successfully distinguish the disease state of tissues and determine the presence and extent of tumour [2], so that interaction processes on the scale of a few milli-electron volts must also be occurring in organisms of a higher level of complexity. The important question of this section thus becomes: can the applied THz EMR associated with new technological applications perturb such specific processes, thereby causing damage?

It is appropriate to make some general observations that broadly follow from the over-arching biological principle of *natural selection*. Since THz EMR is naturally screened out by water in the atmosphere, living cells and organisms have not needed to evolve natural defences to this radiation; so, in consequence, it may be possible that a small dose might have a marked effect on living matter. Of course, the entities concerned are constantly irradiated with low levels of EMR from ambient, thermal noise signals: thus they will, by the same principle of natural selection, have evolved to withstand any damage from such local sources; but it may be important to know, or at least estimate, the size of the THz electric fields concerned so that the possibly harmful effects of external irradiation can be gauged. Adair [51, 60, 61] provides a classical discussion of this matter, modelling various biological components (e.g. cell membranes, microtubules) with simple geometric structures, noting characteristic timescales and determining the energy transfer from the noise field into the component under consideration. Natural selection also raises the intriguing question: if an organism could evolve to withstand THz radiation exposure, then what would be the benefits to the organism of such an evolutionary step? [62]. These authors also point out that natural selection has already produced a range of organs in reptiles, insects and other organisms that are receptive to near infrared radiation (2–12 μm); and recently, it has been suggested [63, 64] that sweat ducts and other surface structures in humans and other mammals could be considered as absorption sites for millimetre waves. This is an intriguing topic and the evolutionary consequences for such developed properties warrant closer examination.

A related concern here is the ubiquitous presence of water in, and around, live biological materials: recently, it has been demonstrated that water plays a very significant role in the process of protein folding [65]; and, given the strong absorption of THz EMR by water, it might be expected that protein structure, and therefore function, could be modified by even small amounts of THz EMR. Clearly, given the vast differences in the complexity and structure of biological materials, the sensitivities of materials to THz EMR will vary significantly; but in all cases, water is likely to be the main repository of the absorbed photon energy. For human subjects, THz radiation will not penetrate much further than the skin and so it becomes very important to investigate exposure effects on skin or skin models.

This section is organised as follows: a critical, albeit shortened, summary of previously reported major studies of THz EMR exposure is first presented for entities of various levels of organisation; and a broad overview of the general interaction and dissipation mechanisms is then given, distinguishing between thermal and non-thermal effects. The background experimental investigations that have led to current "legal" exposure standards are then reviewed, focussing on tissue damage and genetic effects: this discussion is complemented by a consideration of some of the difficulties attendant upon achieving such standards. Following this, the elements of the theory used to predict DNA "unzipping" are outlined in more detail, together with a discussion of the related Fröhlich hypothesis. Finally, further experimental studies are suggested and some general conclusions are drawn.

9.7.2 Summary of Major Experimental Data

Over many years there have been persistent reports in the literature that THz irradiation of plants, cells and tissues with millimetre wave or THz EMR may have an effect on: growth rates in plants [66] and yeast [67], wound healing [68] or even complex phenomena such as anxiety levels in small mammals [69]. From the perspective of determining if THz EMR is, or is not, safe for use in surveillance and security systems, the crucially important question is: do such processes damage biological materials, and—if so—what are the conditions of irradiance, wavelength, exposure time and (if appropriate) pulse structure that should be avoided?

The first investigations of potentially harmful properties of THz EMR were reported in 1968, during studies of cell dehydration, when changes in metabolic rates were observed for cell irradiation at 0.136 THz [70]. Catalogues of millimetre wave exposure effects have been provided by the reviews of Pakhomov et al. [71] and for lower frequency THz irradiation by Fedorov et al. [72]. Very comprehensive and authoritative recent reviews are by Ramundo-Orlando and Gallerano and also by Wilmink and Grundt [73, 74]. As mentioned previously, an exceptionally diverse range of effects are noted in these reviews, across organisms in both the animal and plant kingdoms and in cells, organelles and biomolecules. It is almost impossible to draw out any common themes from these wide observations, but the following very general comments are offered:

- In some investigations, there appear to be marked responses occurring at specific wavelengths. These are perhaps attributable to the absorption of THz EMR changing the activity of a key molecule such as nitric oxide (NO) [75–77] or the direct promotion of a bio-molecular reaction [78, 79] leading to mutation or some other effect.
- In some exposure and irradiation studies [80, 81] made using pulsed THz EMR, importance attaches to the repetition rate of the pulses. The possible implication of this finding is that the THz EMR is directly inducing a change in electrical potential (that might encourage or inhibit a process such as neuron firing) or stimulating a mechanical motion within a cell membrane where there is a natural resonance with the repetition rate.
- Whilst exposure effects vary markedly, there is some indication that a time threshold exists for the production of biological effects [49, 82]. Therefore, models for

non-thermal effects (see below) should be capable of explaining such phenomena, perhaps in terms of the time required for the build-up of an internal energy state [83].

• As noted by the above-quoted reviewers, and elsewhere [84], THz radiation of sufficient intensity can cause irreversible and permanent damage to cells and tissues through temperature increases. Moreover, the cell growth may be increased under certain exposure conditions [85–87].

For comprehensive and clear tabular summaries of the effects of THz exposure studies reported to date on tissues, organisms, cells and bio-molecules, the reader is referred to [74] and to [84]. It is immediately apparent from these tables that conditions of exposure time, irradiance, wavelength, pulse structure (if any) and sample temperature control are extremely varied across the investigations reported and a much more co-ordinated programme of research is required to fully explore the phenomena and explain their physical origins.

9.7.3 Interaction Mechanisms for THz Frequency EMR and Biological Materials

It is convenient to classify the possible effects of THz EMR on biological entities of varying degrees of size and organisation as either *thermal* or *non-thermal* (or: *athermal*, or even *microthermal*) in nature [88]. This separation is somewhat artificial, but it is intended to distinguish those effects where a temperature change is important (e.g. where such changes result in altered biochemical reaction rates) from other effects where the primary interaction involves the electric (or magnetic) field applied to the material. All interactions will, of course, result in energy transfer, leading eventually to temperature changes; and so the timescales for interaction and dissipation (as well as the size of the energies involved) are crucial in determining the relative significance of the two classes of mechanism.

Wilmink and Grundt [74], in their recent extensive review, also make this distinction and indeed go on to further classify phenomena according to biological scale (tissue, cell, organelle, bio-macromolecule). They describe a particularly useful classical modelling approach, in which a variation of the *Pennes' bioheat equation* [89] is solved by either Monte Carlo or Finite Difference Time Domain methods to model the propagation and dissipation of energy. The method has also been used to determine damage thresholds of animal tissue (used to emulate human skin) exposed to a THz beam: the *Pennes* equation is initially solved and a further equation (the *Arrhenius* equation) is then used to predict temperature differences [90]. These authors note that thermal effects on organisms and tissues can produce reversible and irreversible damage that includes: inflammatory response; tissue desiccation and necrosis; and the denaturation of structural protein such as fibrillar collagen. At the cellular level, thermal effects may lead to: changes of cell growth and metabolism; activation of stress response mechanisms; or eventual cell death via apoptosis or necrosis. The important point to note in all such cases is that EMR absorption leading to relatively small (a few degrees Celsius) temperature changes for periods of only several tens of minutes may result in such serious damage. Even more sensitive to thermal effects will be the cellular organelles, through changes to the membranes surrounding them [91] and to the nuclear matrix which displays great sensitivity. Mechanisms such as protein folding and enzyme activity, occurring at the molecular level through changes in hydrogen or other bonding, will all of course be susceptible to thermal effects. These, in turn, may result in either reversible or irreversible damage at the higher level of organisation, dependent on the nature and extent of the thermal shock involved.

The possibility that biological systems may also exhibit athermal changes, when exposed to THz EMR, has been discussed extensively [49, 92, 93]. The experimental evidence for these remains, at best, rather inconclusive although a number of geno-toxic phenomena [49, 50, 85, 92, 94, 95] do seem to be best interpreted using these ideas; further discussion is given below. Central to this discussion is the hypothesis that the interaction of vibrational energy states in living matter leads to an energy "band" in a fashion analogous to electronic bands formed in a semiconductor material through the interaction of the electronic states of individual atoms. This hypothesis, developed by Fröhlich in a series of frequently quoted, but seldom read, papers suggests that such long-lived states (in the "band") can store energy for long times (tens of picoseconds, or longer) resulting in unbinding of the (double-stranded) DNA. Adair [51] discusses limits to such athermal effects by evaluating, on a semi-classical basis, the magnitude of energy transfer for representative biomolecular processes such as electrostriction of cells, linear and rotational biomolecular motion, the opening and closing of voltage-gated channels and the mechanical vibration of cell membranes. Although dealing with rather lower energy EMR, Adair concludes that such athermal effects are unlikely as the 'vibratory motion' involved will be damped by surrounding fluids (especially water) and 'significant energy transfers' will not take place. Similar conclusions are offered by Swanson [96] for THz EMR, as will be discussed later.

9.7.4 Exposure Standards, Supporting Data and Experimental Limitations

The THz region stands at the intersection of the optical and microwave regimes of the electromagnetic spectrum. In consequence, there are two starting points in the determination of safe working levels of THz irradiation and only recently [97, 98] has a coherent view emerged. The original guidelines, prepared by the American National Standards Institute (ANSI) [99], the Institution of Electrical and Electronics Engineers (IEEE) [58] and the International Commission on Non-Ionizing Radiation Protection (ICNIRP) [100], were discussed in some detail first by Berry et al. [101] who drew attention to initial plans for the harmonisation of standards [102] and to the need to take account of the pulsed nature of the exposure when evaluating the safety levels of widely used time domain spectroscopy (TDS) [103] systems. The primary measurement unit in this field is the Maximum Permitted Exposure (MPE). It is important to note that the MPE is regarded as a standard only: exposure to doses of radiation in excess of the MPE is not necessarily harmful. However, in the absence of any clear beneficial effects, exposures to levels greater than the MPE are not recommended.

The above-quoted authors did not conduct any experimental studies but relied instead on extrapolations from the neighbouring MMW and infrared regions in arriving at safe exposure levels. The first satisfactory attempts to determine tissue damage thresholds have recently been reported by Dalzell et al. [90], using pig skin and wet chamois cloth as models for human skin exposed to free electron laser and molecular gas laser THz sources. Their methodology is comprehensive, and included computational modelling of temperature rises and assessments of damage levels undertaken by visual inspection and subsequent statistical analysis. Their measured damage threshold is $7.16\,\mathrm{Wcm^{-2}}$ for THz EMR between 0.1 and 10 THz, a result that is in surprisingly good agreement with model predictions. This threshold is somewhat greater than the conservative MPE values of $1.19\,\mathrm{Wcm^{-2}}$ stipulated by the ICNIRP guidelines [97] or $0.3\,\mathrm{Wcm^{-2}}$ recommended by the EU [98]. Using any of these rules, however, it is clear that TDS systems are quite safe: for such systems, THz power levels of a few microwatts are delivered into a beam that may be a few millimetres in diameter. However, the power levels currently available from QCLs (~hundreds of milliwatts at THz frequency for pulsed sources) do now begin to raise safety concerns [104].

As noted in Sect. 9.7.2, exposure of cells to THz radiation may result in anomalous growth rates and genetic instability as well as permanent and irreversible cell death. In addition to being aware, therefore, of macroscopic damage levels of the type discussed in the present subsection it is important to determine the viability and activity of cells post-exposure and, if possible, to determine the mechanisms for any observed changes. The first such measurements [85] found that low power exposures $(0.03\,\mathrm{mWcm^{-2}})$ could produce genotoxic effects in mammalian cells under long exposures of six hours or more, whereas exposure studies for shorter periods, albeit at a greater irradiance and a different frequency, lead to no apparent chromosomal damage [86]. The first exposure investigations on keratinocytes using a broadband THz source [87] also concluded that cell differentiation was unaffected. Following the observation that microwave radiation could induce hybridisation of DNA [93] at temperatures below the thermal melting temperature, rather precise THz dosimetric studies of post-exposure cell viability, gene-expression profiles and death thresholds were reported using adult human dermal fibroblast and Jurkat cell lines [84, 90, 105]. The major conclusion of these findings is that the biological effects resulting from THz exposure are largely photo-thermal: cells exposed to hyper-thermic stress and THz radiation show comparable levels of gene expression; THz radiation does not induce intracellular DNA repair processes; there appears to be no preferred pathway to cell death (via apoptosis or necrosis); and that the primary genes induced by THz radiation are inflammatory cytokines (which might, in fact, be used as *biomarkers* for THz exposures.) A recent investigation into the specific changes induced by SMMW EMR (106 GHz) does, however, contradict these conclusions and indicates that so-called *spindle disturbances* (which may have implications for disease) can be induced during post-exposure cell division as a direct consequence [106].

In a recent paper [95] attention was turned to the role of THz EMR in promoting the up-regulation of genes that are known to encode for proteins involved with transcription activation, inflammation and cell growth regulation. In a further article by the same authors [93] the possibility was raised that THz radiation may preferentially activate plasma membrane genes and intracellular transduction pathways. The common theme of both papers is that THz radiation may affect signalling mechanisms via a non-thermal mechanism. It is noteworthy that cell-signalling mechanisms that involve conformational changes have previously been proposed on thermodynamics grounds [107]: THz radiation is likely to be sensitive to such changes and might therefore be a suitable probe of cell signalling mechanisms. The authors also tentatively note that "THz radiation may elicit effects that are not fully attributable to the temperature rise provided by the THz radiation". This conclusion is in line with two recent studies [49, 50] where the cellular response of mesenchymal mouse stem cells exposed to THz radiation was investigated. Low-power radiation from both a pulsed broad-band (centred at 10 THz) source and from a CW laser (2.52 THz) source was applied. Using a combination of modelling, empirical characterization and monitoring techniques the impact of radiation-induced increases in temperature was minimized. It was found that the differential expression of the investigated heat shock proteins (HSP105, HSP90, and CPR) was unaffected, while the expression of certain other genes (Adiponectin, GLUT4, and PPARG) showed clear effects of the THz irradiation after prolonged, broad-band exposure. The considerable implications of these findings will be discussed in the next subsection.

To conclude, it is worth remarking that particular attention must be paid in these investigations to all aspects of experimental design. At present, THz power measurements are not normally made with the standard of precision that may be found in other parts of the electromagnetic spectrum. The use of THz power meters that can be [108] referred to national standards laboratories is extremely important, as is the monitoring of cell temperature using fibre-fed thermometers or similar devices. Correct environmental chamber design is also important, together with a good understanding of the beam profile and frequency content. There appears at present to be only one reported study of a simulation of conditions within a sample container irradiated by THz EMR; Jastrow et al. [109] describe a Finite Difference Time Domain simulation of heat distribution and the specific absorption rate in a layer of cells contained within a conventional sample dish placed in an incubator unit exposed to THz radiation. Further design studies of this nature are essential to ensure that meaningful dosimetric data are obtained.

9.7.5 DNA "Unzipping" and the Fröhlich Hypothesis

As is apparent from the previous subsection, there is now emerging a body of evidence to suggest that exposure of living cells to THz radiation can induce a number of non-thermal effects (e.g. under- or over-expression of certain genes). These occur at irradiance levels significantly below the "legal" limits of $5–10\,\text{mW cm}^{-2}$, which have been established from observations of damage levels to tissues mimicking human skin. However, these subtle genetic phenomena do not always appear, but are dependent on a variety of parameters that include: incident power, THz frequency, pulse architecture and exposure times. This is clearly a matter of considerable commercial concern and much more experimentation is needed to establish the genuine extent of the problems and the nature of the mechanisms involved.

The most likely cause of the gene instabilities [49, 50, 85] is a *cooperative* (alternatively: *collective* or *coherent*) effect, first proposed by Fröhlich on purely mathematical grounds [53], which can best be understood using a relatively simple model of DNA dynamics [110] developed by Dauxois, Peyrard and Bishop (DPB). The ideas initially hypothesised by Fröhlich will first be summarised, together with some supporting experimental evidence; and then a review of the important features of this dynamical model of DNA will be presented. It will be seen that the concept proposed by Fröhlich, of a long-lived metastable state in living matter, has a physical embodiment in a specific dynamical mode known as a "breather" which lasts up to 100ps or more, extends over a few angstroms, and is a direct consequence of: the disorder resulting from the sequencing of the DNA base pairs, entropy, nonlinear interactions amongst dynamical modes and anharmonic molecular potentials.

Fröhlich first proposed the idea that living matter might sustain long-lived metastable states, energised by metabolic processes, solely on abstract mathematical grounds. Central to his argument was the concept of nonlinear resonance effects, whereby nonlinear coupling between a range of modes (e.g. vibrational, librational, torsional modes of molecular structures such as DNA) would ensure that the excitation of one mode would be imparted to other modes. An equivalent picture is of the formation of an *energy band*, which leads on to the possibility of coherent effects such as Bose-Einstein condensation [111]. Subsequent papers by Fröhlich and others developed these ideas using more concrete biological examples. These have included: an explanation of enzyme activity [111] the growth of yeast cells [112]; cell-signalling and interaction effects, with particular reference to cancer formation [113, 114] cell membrane dynamics [115] and microtubule dynamics as the origin of "consciousness" [116]. Important developments have been provided by Wu and Austin [83] who provided an expansion of the energy-storage ideas and a helpful explanation for the time-threshold that is observed in radiation-induced effects. Adair [51, 60, 61] provides strong counter arguments to those of Fröhlich, using a classical basis, arguing that any coherent effects that may occur in biological systems are simply damped down by rapid dissipation of energy to water which always surrounds the biological entity concerned. Reimers et al. [117] provide a critical re-appraisal of Fröhlich's condensation ideas, developing a classification of regimes. They conclude that the class of condensation suggested to be responsible for anomalous biological properties, as described in this section, is most unlikely to be encountered in nature. Reimers et al. do, however, state that a variant of the Fröhlich condensation process, as originally envisaged, may indeed occur when biological systems are irradiated with intense THz frequency EMR.

It was hypothesised [59] that large, thermally induced openings (local denaturation; or, *bubbles*) could move slowly along the double strands of DNA (dsDNA); and that these bubbles coincide with the location of the starting-sites for the transcription process. In addition, it is suggested that these bubbles may be associated with the "breathers" or long-lived breathing modes of base pairs [118]. They are, essentially, examples of the Fröhlich condensate, arising from nonlinear interactions of dynamic modes. It is also suggested that they occur at the highly sensitive sites which are rendered unstable by the application of external THz EMR, thus giving rise to the anomalous genetic instability phenomena reported earlier.

The starting point for this discussion is the DPB model of the dynamics of ds-DNA [110] which deploys a Morse potential between base pairs and an anharmonic stacking potential between nearest neighbour bases. This model predicts a co-operative effect, arising from the nonlinear interaction of the phonons, resulting in a much lower than expected melting temperature. Chitanavis [119] develops the DPB model by adding a driving term to describe an externally applied oscillating electric field. Chitanavis also recognizes explicitly the presence of the "solvent" (water) surrounding the dsDNA. Chitanavis concludes that "microwaves of sufficient power…overcome the dissipative effects of the ambient medium…" which would otherwise act to ameliorate the growth of breather modes. (This is a less damning conclusion than that reached by Adair [51, 60, 61], on classical grounds.) Alexandrov et al. [92] extend the DPB model and include a driving term to represent an externally applied THz frequency electric field. They probe the conditions of power, exposure time and EMR frequency under which bubble formation occurs and so genetic instability effects may be expected. These workers conclude that the amplitudes of large, localised openings in the presence of the THz field are significant as the applied EMR resonantly can influence the dynamical stability of the dsDNA, resulting in interference with DNA function. Their treatment provides a feasible mechanism (that is "probabilistic and not deterministic") to explain why long exposures, especially with weak fields, are needed before genetic instabilities develop. Experimental studies that draw on this theoretical model have been reported by Bock et al. [49] and Alexandrov et al. [50]. Bock et al. used a novel broadband THz source [120] with output centred at $10\,THz$, giving an average irradiance of $1\,mW\,cm^{-2}$ and pulse powers with a $30\,MW$ peak. Morphological changes in mouse stem cells were observed after THz irradiation and, although 89% of the cells showed behaviour that did not differ from the (unexposed) control samples, a significant percentage (6%) of the genes on the array were found to increase the level of RNA transcripts, whereas 5% showed a diminished activity level. Particular care was taken in these experiments to ensure that the temperature of the exposed and control cell dishes remained identical to within $\pm 0.2\,°C$. Alexandrov et al. [50] confirm these results and specifically investigate the possibility of cellular stress induced by thermal effects, finding that these are not significant. The overall conclusion is that the response to THz radiation interacts in a sequence-specific manner with the DNA and that this might be harnessed to manipulate and control cell re-programming in a productive manner.

The most recent contribution to this debate critically examines the extensions of the DPB theory proposed by Alexandrov et al. [50] to explain the above data. Swanson [96] argues that "it is extremely unlikely" that the dsDNA denaturing process could be induced by normally accessible sources of THz radiation. The argument is based on the sensitive parameter dependence of the length of time that the breather modes are stable. By making slight changes in these parameters (describing the molecular potentials, the damping and the drive amplitude and frequency) it is shown that the destabilisation time would change from more than 100 ps to less than 20 ps. Swanson further contends that the physical assumptions underlying the DPB model are inaccurate: this author conducts a very simple classical analysis of the force that can be exerted on a singly charged nucleotide in the THz field that corresponds to the "legal" maximum of approximately $10\,\mathrm{mW\,cm^{-2}}$. He concludes that such a field delivers a driving force of around 10^{-24} N, whereas calculations suggest that a force of around 100 pN or greater is needed to create the breather mode. The final point made parallels the earlier arguments from natural selection: that evolution will have inevitably produced a molecule that is stable to non-ionizing radiation and hence immune to low intensity THz EMR.

9.7.6 Conclusions: Is THz EMR Safe?

This section has described a variety of experimental studies that have been conducted to determine if THz EMR is, or is not, safe for biological entities of various levels of organization. At first sight it would seem that THz EMR, as a non-ionizing radiation, must be intrinsically safe although it is known that a variety of biological processes involve energy scales that are characteristic of the THz frequency band. A considerable number of effects have been reported over more than three decades that suggest THz radiation can produce unusual effects such as growth enhancements, wound healing or changes in anxiety levels.

Although initially there were no clear standards for THz exposure promulgated by international standards bodies, recent harmonisation of standards across the millimetre wave and infrared bands implied exposure levels on the order of a few milliwatts per square centimetre. The most careful and recent investigations, reinforced by simulations, suggest that this is somewhat conservative and that tissue damage is unlikely to occur below $7–10\,\mathrm{mWcm^{-2}}$. This indicates that time domain spectroscopy systems are essentially safe, but some caution should be exercised with prolonged exposure of THz EMR from some types of QCLs.

The general interaction mechanisms between THz EMR and biological materials have been reviewed, together with the most recent studies of THz exposure of tissues and cells. Great care has been taken in many of these investigations to ensure that temperature changes in the samples under test do not exceed those in control samples by more than a very small amount, typically $\pm 0.2°$. Although there is still some disagreement, the consensus view is that some genetic instability may be induced in a non-thermal manner under specific conditions of high intensity and long exposure. The theoretical rationale for this interaction involves the formation of long-lived, localised, breather modes in dsDNA which might also be viewed as examples of the metastable states in living matter first hypothesised in the late 1960s.

More experimentation is now needed to determine unequivocally the radiation parameters and exposure lengths, the genetic sequences and the effects of the surrounding solvent (water) that produce damage. Of particular importance will be the use of high power and extremely short (ps) THz pulses which can encourage dynamical breather growth, before the energy transferred from the EMR is rapidly dissipated away to the solvent phonon bath. Given this comment, it may indeed be imprudent to expose human subjects (including operating personnel) to THz security and surveillance systems for extended periods of time.

9.8 Overall Conclusions

The main conclusions for security systems are:

- THz spectroscopy is probably a "bridge too far" and now seems unlikely for stand-off security scanning.
- High frequency radar appears promising for long range stand-off detection.
- Ethics and the safety of security systems are now being properly considered.

The main conclusions for safety considerations are:

- Current TDS systems, where the power level is a few μw in a beam diameter of 5 mm, are well below safety thresholds predicted by several methods.
- QCLs which produce 100 mW or so are potentially more dangerous and care should be exercised in any THz system that uses these routinely.
- In determining the MPE for THz frequencies, it is noted that there are significant problems with measuring incident power levels which should ideally be traceable to national standards. Furthermore, attention also needs to be paid to properly characterising the THz beam and carefully monitoring temperature changes in experimental biological samples.
- THz MPE safety standards of around $1-3\,\mathrm{mWcm^{-2}}$ are conservative in setting a level at which tissue damage may occur; and recent work suggests that this could be safely increased to $7-10\,\mathrm{mWcm^{-2}}$.
- Although there is still some disagreement, the consensus view is that some genetic instability may be induced in a non-thermal manner under specific conditions of high intensity and long exposure. The theoretical rationale for these phenomena involves the formation of long-lived, localised, breather modes in dsDNA which might also be viewed as examples of the metastable states in living matter first hypothesised in the late 1960s.

References

1. C.D. Stoik, M.J. Bohn, J.L. Blackshire, Nondestructive evaluation of aircraft composites using transmissive THz time domain spectroscopy. Opt. Express **16**, 7039–17051 (2008)
2. E. Pickwell, V.P. Wallace, Biomedical applications of THz technology. J. Phys. D Appl. Phys. **39**, R301–R310 (2006)
3. A.J. Baragwanath, G.P. Swift, D. Dai, A.J. Gallant, J.M. Chamberlain, Silicon based microfluidic cell for THz frequencies. J. Appl. Phys. **108**, 1013102 (2010)
4. H.-J. Song, T. Nagatsuma, Present and future of THz communications. IEEE Trans. THz Sci. Technol. **1**, 256–263 (2011)
5. Terahertz systems: technology and emerging markets (2012), http://www.marketresearch.com/Thintri-Inc-v3207/Terahertz-Systems-Technology-Emerging-2613352/. Accessed 8 Feb 2012
6. R.S.King, How 5 technologies fared after 9/11, IEEE Spectrum,48(9), 13, (2011)
7. R.A. Quinn, A.F. Wolkenstein, B. Rampp, *An Ethics of Body Scanners Requirements and Future Challenges from an Ethical Point of View*, vol. 8022, eds. by A. Luukanen, D.A. Wikner (SPIE, Orlando, 2011), p. 80220Q
8. R. Appleby, H.B. Wallace, Standoff detection of weapons and contraband in the 100 GHz to 1 THz region. IEEE Trans. Antennas Propag. **55**(11), 2944–2956 (2007)
9. K.E. Peiponen, J.A. Zeitler, K. Kuwata-Gonokami, *Terahertz Spectroscopy: Theory and Applications*. (Springer, Berlin, 2011)
10. *Assessment of Millimeter-Wave and Terahertz Technology for Detection and Identification of Concealed Explosives and Weapons*. (The National Academic Press, Washington, 2007)
11. D.L. Woolard, W.R. Loerop, M.S. Shur, *Terahertz Sensing Technology*. (World Scientific, Singapore, 2004)
12. M.C. Kemp, P.F. Taday, B.E. Cole, J.A. Cluff, A.J. Fitzgerald, W.R. Tribe, Security applications of terahertz technology. Proc. SPIE Int. Soc. Opt. Eng. **5070**, 44–52 (2003)
13. D.L. Woolard, E.R. Brown, M. Pepper, M. Kemp, Terahertz frequency sensing and imaging: a time of reckoning future applications? Proc. IEEE **93**(10), 1722–1743 (2005)
14. Kemp, M. C.: Millimetre wave and terahertz technology for the detection of concealed threats: a review, *in Proc. SPIE Int. Soc. Opt. Eng.* (USA) Stockholm, Sweden: SPIE - The International Society for Optical Engineering, **6402** 64020-64021 (2006)
15. Kemp, M. C.: Detecting hidden objects: Security imaging using millimetre-waves and terahertz. in *Proceedings of the 2007 IEEE Conference on Advanced Video and Signal Based Surveillance (AVSS 2007)*. (Institute of Electrical and Electronics Engineers Inc, London, 2007), pp. 7–9 (NJ 08855–1331)
16. M.C. Kemp, Millimetre wave and terahertz technology for detection of concealed threats—a review. in *Proceedings of the 32nd International Conference on Infrared and Millimeter Waves and the 15th International Conference on Terahertz Electronics (IRMMW-THz)*. (IEEE, Cardif, 2008), pp. 647–648
17. M. Kemp, Explosives detection by terahertz spectroscopy—a bridge too far? IEEE Trans. Terahertz Sci. Technol. **1**(1), 282–292 (2011)
18. S.G. Kong, D.H. Wu, TeraHertz time-domain spectroscopy for explosive trace detection, computational intelligence for homeland security and personal safety. in *Proceedings of the 2006 IEEE International Conference on Computational TeraHertz Time-Domain Spectroscopy for Explosive Trace Detection*. (2006), pp. 47–50
19. Y. Dikmelik, M.J. Fitch, M.R. Leahy-Hoppa, R. Osiander, Examining explosive residues on surfaces with terahertz technology. Proc. SPIE Int. Soc. Opt. Eng. **6549**, 65490–65491
20. D.M. Sheen, D.L. McMakin, T.E. Hall, Three-dimensional millimeter-wave imaging for concealed weapon detection. IEEE Trans. Microwave Theory Tech. **49**(9), 1581–1592 (2001)
21. R. Appleby, Passive millimetre-wave imaging and how it differs from terahertz imaging. Philos. Trans. R. Soc. Lond. Ser. A (Math. Phys. Eng. Sci.) **362**(1815), 379–394 (2004)

22. M. Tiebout, H.D. Wohlmuth, H. Knapp, R. Salerno, M. Druml, J. Kaeferboeck, M. Rest, J. Wuertele, S.S. Ahmed, A. Schiessl, R. Juenemann, Low power wideband receiver and transmitter chipset for mm-wave imaging in SiGe bipolar technology, Radio Frequency Integrated Circuits Symposium (RFIC), IEEE (2011) pp. 1–4
23. S.S. Ahmed, A. Schiessl, L. Schmidt, Novel fully electronic active real-time millimeter-wave imaging system based on a planar multistatic sparse array Microwave Symposium Digest (MTT), IEEE MTT-S International, (2011)
24. P. Corredoura, Z. Baharav, B. Taber, G. Lee, Millimeter-wave imaging system for personnel screening: scanning 10^7 points a second and using no moving parts, eds. by R. Appleby. D.A. Winker, vol. 6211 (SPIE, Orlando, 2006), pp. 62110B–8
25. Smiths Detection: eqo Revolutionising Peopel Screening (2011), http://www.smithsdetection.com/eqo.php Accessed 23 Dec 2011
26. K.B. Cooper, R.J. Dengler, N. Llombart, T. Bryllert, G. Chattopadhyay, E. Schlecht, J. Gill, C. Lee, A. Skalare, I. Mehdi, P.H. Siegel, Penetrating 3-D imaging at 4-and 25-m range using a submillimeter-wave radar. IEEE Trans. Microwave Theory Tech. **56**(12), 2771–2778 (2008)
27. K.B. Cooper, R.J. Dengler, N. Llombart, A. Talukder, A.V. Panangadan, C.S. Peay, I. Mehdi, P.H. Siegel, Fast high-resolution terahertz radar imaging at 25 meters, Proc. SPIE Int. Soc. Opt. Eng. **7671** (2010)
28. D.M. Sheen, T.E. Hall, R.H. Severtsen, D.L. McMakin, B.K. Hatchell, P.L.J. Valdez, Standoff concealed weapon detection using a 350-GHz radar imaging, system, Proc. SPIE Int. Soc. Opt. Eng. **7670** (2010)
29. L. Marchese, M. Bolduc, B. Tremblay, M. Doucet, H. Oulachgar, L. Le Noc, F. Williamson, C. Alain, H. Jerominek, A. Bergeron, A microbolometer-based THz imager. Proc. SPIE Int. Soc. Opt. Eng. **7671**, 76710Z (2010)
30. M. Bolduc, L. Marchese, B. Tremblay, M. Doucet, M. Terroux, H. Oulachgar, L. Le Noc, C. Alain, H. Jerominek, A. Bergeron, Video-rate THz imaging using a microbolometer-based camera. in *35th International Conference on Infrared, Millimeter, and Terahertz Waves (IRMMW-THz 2010)*. (IEEE, Piscataway, 2010), p. 2
31. J. Meilhan, B. Dupont, V. Goudon, G. Lasfargues, J. Lalanne Dera, D.T. Nguyen, J.L. Ouvrier-Buffet, S. Pocas, T. Maillou, O. Cathabard, S. Barbieri, F. Simoens, Active THz imaging and explosive detection with uncooled antenna-coupled microbolometer arrays. Proc. SPIE Int. Soc. Opt. Eng. **8023**, 80230E (2011)
32. P. Dean, N.K. Saat, S.P. Khanna, M. Salih, A. Burnett, Dual-frequency imaging using an electrically tunable terahertz quantum cascade laser. Opt. Express **17**(23), 20631–20641 (2009)
33. F. Simoens, A. Arnaud, P. Castelein, V. Goudon, P. Imperinetti, J. Lalanne Dera, J. Meilhan, J.L. Ouvier Buffet, S. Pocas, T. Maillou, L. Hairault, P. Gellie, S. Barbieri, C. Sirtori, Development of uncooled antenna-coupled microbolometer arrays for explosive detection and identification. Proc. SPIE Int. Soc. Opt. Eng. **7837**, 78370B (2010)
34. H.W. Hubers, H. Richter, A.D. Semenov, S.G. Pavlov, L. Mahler, A. Tredicucci, H.E. Beere, D.A. Ritchie, K. Il'in, M. Siegel, Progress towards a 2.5-THz solid state heterodyne receiver with quantum cascade laser and hot electron bolometric mixer, *Infrared, Millimeter and Terahertz Waves, 2008. IRMMW-THz 2008.33rd International Conference on* (2008), p. 1
35. A. Luukanen, L. Gronberg, P. Helisto, J.S. Penttila, H. Seppa, H. Sipola, C.R. Dietlein, E.N. Grossman, Passive Euro-American teraherz camera (PEAT-CAM): passive indoor THz imaging at video rates for security applications. Proc. SPIE Int. Soc. Opt. Eng. **6548**, 654808–1 (2007)
36. A. Luukanen, M. Anholm, P. Lappalainen, M. Leivo, A. Rautiainen, A. Tamminen, J. Ala-Laurinaho, A.V. Raisanen, C.R. Dietlein, E. Grossman, Passive real-time submillimetre-wave imaging system utilizing antenna-coupled microbolometers for stand-off security screening applications, International Workshop on Antenna Technology: "Small Antennas, Innovative Structures and Materials" (iWAT 2010)
37. A. Luukanen, M. Aikio, M. Gronholm, M.M. Leivo, A. Mayra, A. Rautiainen, H. Toivanen, Design and performance of a passive video-rate THz system demonstrator. Proc. SPIE Int. Soc. Opt. Eng. **8022**, 802207 (2011)

38. E.N. Grossman, Broadband and multispectral response of planar antennas for terahertz security screening, in *2011 5th European Conference on Antennas and Propagation (EuCAP)* (IEEE, Piscataway, 2011), pp. 3171–3172

39. E. Heinz, D. Born, G. Zieger, T. May, T. Krause, A. Kruger, M. Schulz, S. Anders, V. Zakosarenko, H.G. Meyer, M. Starkloff, M. Rossler, G. Thorwirth, U. Krause, Progress report on Safe VISITOR: approaching a practical instrument for terahertz security screening. Proc. SPIE Int. Soc. Opt. Eng. **7670**, 767005 (2010)

40. E. Heinz, T. May, D. Born, G. Zeiger, S. Anders, V. Zakosarenko, M. Schubert, T. Krause, A. Kruger, M. Schulz, H.G. Meyer, Towords high-sensitivity and high-resolution submillimeter-wave video imaging. Proc. SPIE Int. Soci. Opt. Eng. **8022** (2011)

41. J.F. Federici, A.M. Sinyukov, I. Zorych, Z. Liu, R.B. Barat, D.E. Gary, Z.H. Michalopoulou, *Terahertz Synthetic Aperture and Interferometric Imaging for Concealed Weapons and Explosives Detection* (NATO, Bucharest, 2008)

42. A. Majewski, R. Abreu, M. Wraback, A high resolution terahertz spectrometer for chemical detection. Proc. SPIE Int. Soc. Opt. Eng. **6549**, 65490–65491 (2007)

43. X. Xu, D. Jianming, M. Yamaguchi, X.C. Zhang, Using air as the nonlinear media for THz wave generation. Proc. SPIE Int. Soc. Opt. Eng. **6212**, 62120–62121 (2006)

44. M.D. Thomson, M. Kreb, T. Loeffler, H.G. Roskos, Broadband THz emission from gas plasmas induced by femtosecond optical pulses: from fundamentals to applications. Laser Photonics Rev. **1**(4), 349–368 (2007)

45. J. Liu, X. Zhang, Remote terahertz wave sensing using laser-induced fluorescence. Phy **39**(6), 419–422 (2010)

46. E. Ojefors, A. Lisauskas, D. Glaab, H.G. Roskos, U.R. Pfeiffer, Terahertz imaging detectors in CMOS technology. J. Infrared Millimeter Terahertz Waves **30**(12), 1269–1280 (2009)

47. F. Schuster, D. Coquillat, H. Videlier, M. Sakowicz, F. Teppe, L. Dussopt, B. Giffard, Broadband terahertz imaging with highly sensitive silicon CMOS detectors. Opt. Express **19**(8), 7827–7832 (2011)

48. J.W. May, G.M. Rebeiz, High-performance W-band SiGe RFICs for passive millimeter-wave imaging. in *IEEE Radio Frequency Integrated Circuits Symposium.* (IEEE, 2009), pp. 437–440

49. J. Bock, Y. Fukuyo, S. Kang, M.L. Phipps, L.B. Alexandrov, K. Rasmussen, A.R. Bishop, E.D. Rosen, J.S. Martinez, T. Chen, G. Rodriguez, B.S. Alexandrov, A. Usheva, Mammalian Stem Cells Reprogramming in Response to THz Radiation. PLoS. ONE **5**(12), 15806–15815 (2010)

50. B.S. Alexandrov, K. Rasmussen, A.R. Bishop, A. Usheva, L.B. Alexandrov, S. Chong, Y. Dagon, L.G. Booshehri, C.H. Mielke, M.L. Phipps, J.S. Martinez, T. Chen, G. Rodriguez, Non-thermal effects of THz radiation on gene expression in mouse stem cells. Biomed. Opt. Express **2**(9), 2679–2689 (2011)

51. R.K. Adair, Biophysical limits on athermal effects of RF and microwave radiation. Bioelectromagnetics **24**, 39–48 (2003)

52. M. Swicord, E. Postow, Modulated fields and window effects. In: C. Polk, E. Postow, eds. *CRC Handbook of Biological Effects of Electromagnetic Fields and Cells.* vol. 335 (Plenum Publishing Co, New York, 1996)

53. H. Fröhlich, Long lived coherence and energy storage in biological systems. Int. J. Quantum Chem. **2**(5), 641–649 (1968)

54. R. Penrose, *The Emperor's New Mind: Concerning Computers, Minds, and the Laws of Physics.* (Oxford University Press, Oxford, 1999)

55. M.J. Sutcliffe, L. Masgrau, A. Roujeinikova, L.O. Johannissen, P. Hothi, J. Basran, K.E. Ranaghan, A.J. Mulholland, D. Leys, N.S. Scrutton, Hydrogen tunnelling in enzyme catalyzed H-transfer reactions: flavoprotein and quinoprotein systems. Phil. Trans. Roy. Soc. London B **361**, 1375–1386 (2006)

56. B.M. Fischer, M. Walther, P.U. Jepsen, Far-infrared vibrational modes of DNA components studied by THz time-domain spectroscopy. Phys. Med. Biol **47**, 3807–3813 (2002)

57. G.I. Groma, J. Hebling, I.Z. Kozma, G. Varo, J. Hauer, J. Kuhl, E. Riedle, THz radiation from bacteriorhodopsin reveals correlated primary electron and proton transfer process. PNAS **105**, 6888–6893 (2008)
58. *Standards for Safety Levels with Respect to Human Exposure to Radiofrequency Electromagnetic Fields, 3kHz to 300GHz.* (Institute of Electrical and Electronic Engineers, New York, 1999)
59. G. Kaloskas, K. Rasmussen, A.R. Bishop, C.H. Choi, A. Usheva, Sequence specific thermal fluctuations identify start sites for DNA transcription. Europhys. Lett. **68**(1), 127–133 (2004)
60. R.K. Adair, Constraints on biological effects of weak extremely low frequency electromagnetic fields. Phys. Rev. A **43**(2), 1039–1048 (1991)
61. R.K. Adair, Vibrational resonances in biological systems at microwave frequencies. Biophys. J. **82**, 1147–1152 (2002)
62. R. Shiurba, T. Hiyabayashi, M. Masuda, A. Kawamura, Y. Komoike, W. Klitz, K. Kinowaki, T. Funatsu, S. Kondo, S. Kiyokawa, T. Sugai, K. Kawamura, H. Namiki, T. Higashinakagawa, Cellular responses of the ciliate, Tetrahymena thermophila, to far infrared irradiation. Photoch. Photobio. Sci. **5**, 799–807 (2006)
63. G. Shafirstein, E.G. Moros, Modelling millimetre wave propagation and absorption in a high resolution skin model: the effect of sweat glands. Phys. Med. Biol. **56**, 1329–1339 (2011)
64. Y. Feldman, A. Puzenko, P.B. Ishai, A. Caduff, A.J. Agranat, Human skin as arrays of helical antennas in the millimeter and submillimeter wave range. Phys. Rev. Lett. **100**, 1329–1339 (2008)
65. U. Heugen, G. Schwabb, E. Brundermann, M. Hyden, X. Yu, D.M. Leitner, M. Havenith, Solute-induced retardation of water dynamics probed directly by THz spectroscopy. PNAS **103**, 12301–12306 (2006)
66. N.G. Shestopalova, B.I. Makarenko, L.N. Golovina, V.S. Miroschnichenko, Modification of synchronizing effects on first mitoses by different temperature regimes of germination. in *Proceedings of the 10th Russian Symposium on Millimeter Waves in Medicine and Biology.* (1995), pp. 236–237
67. S. Hadjiloucas, M. Chahal, J. Bowen, Preliminary results on the non-thermal effects of GHz radiation on the growth rate of S. cerevisiae cells in microcolonies. Phys. Med. Biol. **47**(21), 3831–3839 (2002)
68. N.N. Korpan, T. Saradeth, Clinical effects of continuous microwave exposure for postoperative sceptic wound treatment. Am. J. Surg. **170**, 271–276 (1995)
69. N. Bondar, I.L. Kovalenko, D.F. Avgustinovich, A.G. Khamoyan, N.N. Kudryavtseva, Behavioral effect of THz waves in male mice. Bull. Exp. Biol. Med. **145**, 401–405 (2008)
70. S.J. Webb, D.D. Dodds, Inhibition of bacterial cell growth by 136 gc microwaves. Nature **218**, 374–375 (1968)
71. A.G. Pakhomov, Y. Akyel, O.N. Pakhomova, B.E. Stuck, M.R. Murphy, Current state and implications of research on biological effects of millimeter waves. Bioelectromagnetics **9**, 393–413 (1998)
72. V.I. Fedorov, S.S. Popova, A.N. Pisarchik, Dynamic effects of submillimeter wave radiation on biological objects of various levels of organization. Int. J. Infrared Millimeter Waves **24**, 1235–1254 (2003)
73. A. Ramundo-Orlando, G.P. Gallerano, THz radiation effects and biological applications. J. Infrared Millimeter THz Waves **30**, 1308–1318 (2009)
74. G.J. Wilmink, J.E. Grundt, Current state of research on biological effects of THz radiation. J. Infrared Millimeter THz Waves **32**, 1074–1122 (2011)
75. V.F. Kirikchuk, N.V. Effimova, E.V. Andronov, Effect of high power THz irradiation on platelet aggregation and behavioral reactions of albino rats. Bull. Exp. Biol. Med. **148**(5), 746–749 (2009)
76. V.F. Kirikchuk, A.N. Ivanov, E.G. Kulapina, A.P. Krenickiy, A.V. Mayborodin, Effect of THz electromagnetic radiation at nitric oxide frequencies on concentration of nitrites in blood serum of albino rats under conditions of immobilization stress. Bull. Exp. Biol. Med. **149**(2), 174–176 (2010)

77. V.F. Kirikchuk, A.N. Ivanov, T.S. Kirijazi, Correction of microcirculatory disturbances with THz electromagnetic radiation at nitric oxide frequencies in albino rats under conditions of acute stress. Bull. Exp. Biol. Med. **151**(3), 288–291 (2011)
78. V.M. Govorun, V.E. Tretiakov, N.N. Tulyakov, V.B. Fleurov, A.I. Demin, A. Yu, V.A. Batanov, A.B. Kapitanov, Far-infrared radiation effect on the structure and properties of proteins. Int. J. Infrared Millimeter Waves **12**(2), 1469–1474 (1991)
79. A. Homenko, B. Kapilevich, R. Kornstein, M.A. Firer, Effects of 100 GHz radiation on alkaline phosphatase activity and antigen-antibody interaction. Bioelectromagnetics **30**, 167–175 (2009)
80. P.H. Siegel, V. Pikov, Can neurons sense millimeter waves?. in *Proceedings of the IEEE 35th International Conference on Infrared, Millimeter and THz Wave.* (2010)
81. A. Ramundo-Orlando, G.P. Gallerano, P. Stano, A. Doria, E. Giovenale, G. Messina, M. Cappelli, M. D'Arienzo, I. Spassovsky, Permeability changes induced by 130 GHz pulsed radiation on cationic liposomes loaded with carbonic anhydrase. Bioelectromagnetics **22**, 303–313 (2007)
82. A.Z. Smolanskaya, R.L. Vilenskaya, Effects of millimeter-band electromagnetic radiation on the functional activity of certain genetic elements of bacterial cells. Soviet. Phys. Uspekhi. **6**, 303–313 (1974)
83. T.M. Wu, S.J. Austin, Fröhlich's model of Bose condensation in biological systems. J. Biol. Phys. **9**, 97–107 (1981)
84. G.J. Wilmink, B.D. Rivest, C.C. Roth, B.L. Ibey, J.A. Payne, L.X. Cundin, J.E. Grundt, X. Peralta, D.G. Mixon, W.P. Roach, In vitro investigation of the biological effects associated with human dermal fibroblasts exposed to 2.52 THz radiation. Lasers Surg. Med. **3**, 152–163 (2011)
85. A. Korenstein-Ilan, A. Barbul, P. Hasin, A. Eliran, A. Gover, R. Korenstein, Terahertz radiation increases genomic instability in human lymphocytes. Radiat. Res. **170**(2), 224–234 (2008)
86. M.R. Scarfi, M. Romano, R. DiPetro, O. Zeni, A. Doria, G.P. Gallerano, E. Giovanele, G. Messina, A. Lai, G. Campurra, D. Coniglio, M. D'Arienzo, THz exposure of whole blood for the study of biological effects on human lymphocytes. J. Biol. Phys. **29**(2), 171–176 (2003)
87. R.H. Clothier, N. Bourne, Effects of THz exposure on human primary keratinocyte differentiation and viability. J. Biol. Phys. **29**(2), 179–185 (2003)
88. L.J. Challis, Mechanisms for interaction between RF fields and biological tissue. Bioelectromagnetics **7**, S98–S106 (2005)
89. P.H. Pennes, Analysis of tissue and arterial blood temperatures in the resting human forearm. J. Appl. Physiol. **1**(2), 93–122 (1948)
90. D.R. Dalzell, J. McQuade, R. Vincelette, B.L. Ibey, J. Payne, R. Thomas, W.P. Roach, C.L. Roth, G.J. Wilmink, *Damage Thresholds for THz Radiation*, eds. by E.D. Jansen, R.J. Thomas, **756** (2010), p. 75620M
91. J.L. Roti Roti, Heat-induced alterations of nuclear protein associations and their effects on DNA repair and replication. Int. J. Hyperthermia **23**(1), 3–15 (2007)
92. B.S. Alexandrov, V. Gelev, A.R. Bishop, A. Usheva, K. Rasmussen, DNA breathing dynamics in the presence of a THz field. Phys. Lett. A **374**, 1214–1217 (2010)
93. W.F. Edwards, D.D. Young, A. Deiters, The effect of microwave irradiation on DNA hybridization. Org. Biomol. Chem. **7**, 2506–2508 (2009)
94. G.J. Wilmink, J.E. Grundt, C. Cema, C.C. Roth, M.A. Kuipers, I. Lipscomb, I. Echchgadda, B.L. Ibey, THz radiation preferentially activates the expression of genes responsible for the regulation of plasma membrane properties. in *Proceedings of the 36th International Conference on Infrared, Millimeter and THz Waves.* (2011)
95. J.E. Grundt, C. Cema, C.C. Roth, B.L. Ibey, I. Lipscomb, I. Echchgadda, G.J. Wilmink,THz radiation triggers a signature gene expression profile in human cells. in *Proceedings of the 36th International Conference on Infrared, Millimeter and THz Waves.* (2011)
96. E.S. Swanson, Modeling DNA response to THz radiation. Phys. Rev. E. **83**, 2506–2508 (2011)
97. J.A. Bain, J.A. Rusch, B.E. Line, ICNIRP statement on far infrared radiation exposure. Health Phys. **91**(6), 630–645 (2006)

98. EC Directive On the minimum health and safety requirements regarding the exposure of workers to risks arising from physical agents (artificial optical radiation). Off J. Eur. Union **114**(1), 38–59 (2006)
99. American national standard for the safe use of lasers: ANSI Z136.1 (2007). (Laser Institute of America, Orlando FL, 2007)
100. International commission on non-ionizing radiation protection (ICNIRP), guidelines on limits of exposure to laser radiation of wavelengths between 180nm and 1.000m. Health Phys. **71**, 804–819 (1996)
101. G.C. Walker, E. Berry, N.N. Zinov'ev, A.J. Fitzgerald, R.E. Miles, J.M. Chamberlain, M.A. Smith, THz imaging and international safety guidelines. Proc. SPIE Int. Soc. Opt. Eng **4682**, 683–690 (2002)
102. International commission on non-ionizing radiation protection, guidelines for limiting exposure to time-varying electric, magnetic and electromagnetic fields up to 300GHz. Health Phys. **74**(4), 494–522 (1988)
103. D. Auston, K. Cheung, P. Smith, Picosecond photoconducting hertzian dipoles. Appl. Phys. Lett. **45**(3), 284–286 (1984)
104. B.S. Williams, S. Kumar, Q. Hu, J.L. Reno, High-power THz quantum-cascade lasers. Electron. Lett. **42**(2), 89–90 (2006)
105. G.J. Wilmink, B.D. Rivest, B.L. Ibey, C.L. Roth, J. Bernhard, W.P. Roach, Quantitative investigation of the bioeffects associated with THz radiation. Proc. SPIE Int. Soc. Opt. Eng **7562**, 89–90 (2010)
106. H. Hintzsche, C. Jastrow, T. Kleine-Ostmann, H. Stopper, E. Schmid, T. Schrader, THz radiation induces spindle disturbances in human-hamster hybrid cells. Radiat. Res. **175**, 569–574 (2011)
107. R. Hawkins, T.C.B. McLeish, Coarse-grained model of entropic allostery. Phys. Rev. Lett. **93**, 569–574 (2004)
108. Absolute terahertz power/energy meter (2012), http://www.THz.co.uk/index.php?option=com_content&view=article&id=140&Itemid=443 Accessed 8 Feb 2012
109. C. Jastrow, T. Kleine-Ostmann, T. Schrader, Numerical dosimetric calculations for in vitro field expositions in the THz frequency range. Adv. Radio Sci. **8**, 1–5 (2010)
110. T. Dauxois, M. Peyrard, A.R. Bishop, Entropy-driven DNA denaturation. Phys. Rev. E **47**(1), R44–R47 (1993)
111. H. Fröhlich, Bose condensation of strongly excited longitudinal electric modes. Phys. Lett. A **26**, 402–403 (1968)
112. W. Grundler, F. Keilmann, H. Fröhlich, Resonant growth rate response of yeast cells irradiated by weak microwaves. Phys. Lett. A **62**(6), 463–466 (1977)
113. H. Fröhlich, Coherent electric vibrations in biological systems and the cancer problem. IEEE Trans Microwave Theory Tech. **26**, 613–617 (1978)
114. H. Fröhlich, Selective long range dissipation forces between large systems. Phys. Lett. A **39**(2), 153–154 (1972)
115. H. Fröhlich, Further evidence for coherent excitations in biological systems. Phys. Lett. A **10**(9), 480–481 (1985)
116. S. Hameroff, Quantum computation in brain microtubules? The Penrose-Hameroff 'ORCH OR' model of consciousness. Phil. Trans. Roy. Soc. Lond. A **356**, 1869–1896 (1988)
117. J.R. Reimers, L.K. McKemmish, R.H. McKenzie, A.E. Mark, N.S. Hush, Weak, strong and coherent regimes of Fröhlich condensation and their applications to THz medicine and quantum consciousness. PNAS **106**(11), 4219–4224 (2009)
118. S. Flach, A.V. Gorbach, Discrete breathers—advances in theory and applications. Phys. Rep. **467**(1–3), 1–116 (2008)
119. S.M. Chitanvis, Can low-power electromagnetic radiation disrupt hydrogen bonds in dsDNA? J. Polym. Sci. Part B **44**, 2740–2747 (2006)
120. K.Y. Kim, A.J. Taylor, J.H. Glownia, G. Rodriguez, Coherent control of THz super-continuum generation in ultrafast laser-gas interactions. Nat. Photonics **2**, 605–609 (2008)